The Ancient Phonograph

The Ancient Phonograph

Shane Butler

ZONE BOOKS · NEW YORK

2015

ZONE BOOKS
633 Vanderbilt Street
Brooklyn, NY 11218

Printed in the United States of America.

Distributed by The MIT Press,
Cambridge, Massachusetts, and London, England.

Library of Congress Cataloging-in-Publication Data

Butler, Shane, 1970–
The ancient phonograph / Shane Butler.
pages cm
Includes bibliographical references and index.
ISBN 978-1-935408-72-7
1. Voice—Philosophy. 2. Voice—Psychological aspects. 3. Classical literature. I. Title.

B105.V64B88 2015
128'.6—dc23

2015004417

Contents

For Jim

But O for the touch of a vanished hand
And the sound of a voice that is still!

—TENNYSON, "BREAK, BREAK, BREAK"

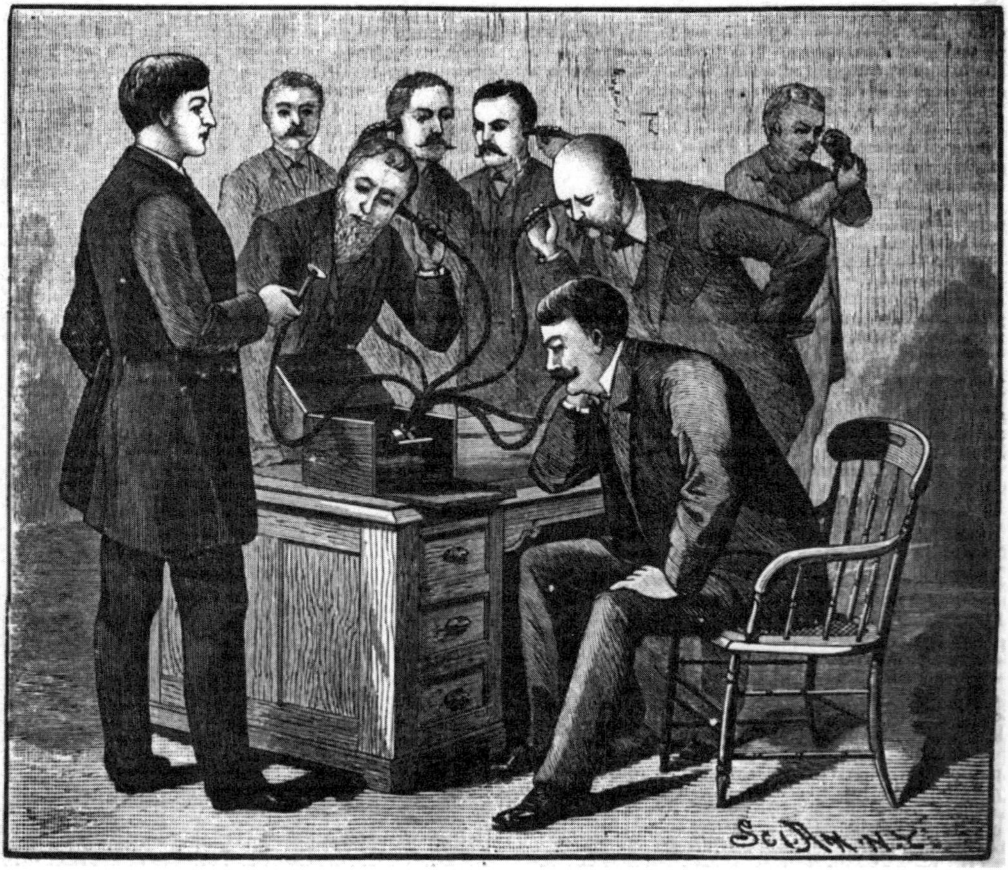

Figure 1. Edison's Phonograph, as first demonstrated to the editors of *Scientific American* in 1877, from "The New Phonograph," *Scientific American Supplement* 632 (1888), 10096.

LINER NOTES

The Ancient Phonograph

> The voice makes people write.
>
> —MICHEL DE CERTEAU, *THE PRACTICE OF EVERYDAY LIFE*

"Good morning. How do you do? How do you like the phonograph?" Thus did Thomas Edison's cheerful new machine greet the editors of *Scientific American*, "to the astonishment of all present."[1] For media theorists, this was one of history's great turning points, and Friedrich Kittler, who begins his account of modern media in Edison's laboratory, looks forward from the recorded salutations of 1877 to a world that, almost at once, would never be the same.[2] The present book looks instead back—indeed, far back, not only before Edison, but long before Marshall McLuhan's earlier starting point of Gutenberg,[3] to an age for which, from our own distant perspective, even writing itself was still relatively new. To return with the right ears to those early chapters of Western writing, I shall argue, is to hear something no less astonishing than what rose from the spinning cylinder of the "wizard of Menlo Park." "How do you like the phonograph?": the new machine's name was a neologism, but like the roots from which that name was compounded, its question was an ancient one.

Etymologically speaking, a "phonograph" proposes to write (*graphein*) the voice (*phōnē*). Edison would monopolize the term but did not invent it, for linguist Edward Hincks had used it earlier in the century to designate those Egyptian hieroglyphs that were "representations of sounds," and the word had entered the general lexicon via an invention that had spread as rapidly as Edison's would: Isaac Pittman's system of shorthand, described in his 1845 *Manual of Phonography, or Writing by Sound.*[4] Pittman and Edison alike thus used the word to announce the arrival of a new kind of writing, but Hincks reveals the subtle problem in this act of branding, for any form of writing that purports to represent speech, such as the ancient syllabic cuneiform he is famous for having deciphered or the alphabetic script in which the words on the present page are written, is in some sense phonographic. Pittman was largely reacting to the vagaries and complexities of English spelling, but Edison's phonographic claim ultimately rests on a far deeper problem, one which, in fact, we can trace back to one of their shared etymological roots, *phōnē*. We shall turn in earnest to this problem in this book's first chapter, but let us anticipate that discussion simply by noting that the Greek word means both "speech" and "voice." If the former meaning makes even the most pedestrian alphabetic texts at least notionally phonographic, in the sense that they inscribe something that has been or could be spoken, the latter leads to conclusions that are far less clear. What did it mean to seek to write the *voice* long before Edison, in faraway Greece and Rome? This question has inspired this book. What has emerged by way of answer is, for its author, a startlingly unfamiliar picture of the aims of ancient writers, striving to capture the voice precisely as something conceptually distinct from language, even if largely inseparable from it. Indeed, the case studies that follow reveal that this voice was more than just a recurring object of desire: rather, it was in antiquity something like the raison d'être of the very category of literature, the texts of which it may even invite us to read as the single experiments of a unified project of phonographic research, stretched out over centuries.

Let us linger for a moment longer among the recorded voices of a more recent past. Early rivals to Edison's phonograph were the "graphophone," from the laboratories of Alexander Graham Bell, and the "gramophone," whose inventor, Emile Berliner, would perfect the flat-disc records that eventually displaced Edisonian cylinders. Though the variation in brand names is partly arbitrary, the inversion of syllables arguably reflects a subtle shift in focus away from the marvelous machine that turned voices into records to the proliferating devices that, in the comfort of countless homes, were turning records back into voices. At the same time, in Berliner's substitution of *gram-* (from the Greek *gramma*, "letter of the alphabet") for *graph-*, we may perhaps detect a look not so much forward as around and back, to mass consumption of the various products of the far earlier invention of the printing press.[5] Novel as it was, the phonograph-gramophone entered bourgeois life in the familiar guise of the fireside reader, a fact from which we may draw two important lessons. First, one cannot go looking for phonographic writing independent of gramophonic reading, for any history of media must also be a history of media players (*lecteurs*, as the French prefer to call them, maintaining the redeployment of writerly language that attended the rise of digital media). Second, the very age that invented the phonograph regarded the gramophonic reanimation of its inscribed voices sufficiently like ordinary reading to market it as such, a strategy that paradoxically captures for us a glimpse of the readerly expectations which Edison's cylinders and Berliner's disks would immediately begin to transform.

These lessons return us to the contradiction that continues to lurk in our own understanding of the relationship between voice and text. While, in one sense, we regularly assume that the voice is indeed what writing captures, especially writing that is "phonetic" (i.e., alphabetic or syllabic), we simultaneously suppose that the voice is precisely that quantity which, before Edison, eluded transcription. We seem to ourselves to resolve this paradox by asserting a distinction between the linguistic voice, which writing has long recorded, and the extralinguistic voice, which had to wait for the phonograph. But centuries of literary texts are filled with—and at least partly defined by—phonic features that cannot be reduced to a function that is, strictly speaking, linguistic, even if we might be inclined to call some of them "expressive" or "communicative." It will be the contention of this book that the ensemble of such features, added to writing's linguistic work, long constituted what we should identify as a phonographic claim. Well over a century of record playing, on a series of machines, has partly deafened us to this claim, even in the case of classical literature, which, as we shall hear, practically shouts it. To be sure, we shall not entirely disagree with Michel de Certeau, for whom the voice is an elusive object of desire that forever propels writing forward. But we shall forgo any post-Edisonian pessimism about the ability of earlier voice-writers to get some satisfaction—or even to provide such now for their acoustically overloaded twenty-first-century readers.

Let us therefore begin again, setting aside more recent phonographs and conjuring that far earlier Edison who, millennia even before Homer, first dazzled his prehistoric contemporaries with a stylus that scratched words into pliant matter—perhaps the same Mesopotamian mud out of which his successors would shape countless cuneiform tablets. Certainly it has been common to assume that, from the start, his aim was to represent spoken language. Strictly speaking, however, we cannot exclude an alternative hypothesis, namely, that his writing sprang, independently, from the same linguistic instinct that had generated human speech; these two forms of language would have been correlated in a second moment (even if this came quickly), through triangulation with their shared ends. Finally, let us imagine an even more radical possibility that detaches writing's origins from any linguistic purpose at all. In this hypothesis, the first writer's aim was the same as Edison's: to capture the voice itself. A grounding in the linguistic voice simply made that task feasible, exploiting speech's existing reduction of the countless variety of sounds human voices can make.

In truth, no such proto-Edison ever existed. Writing developed slowly over millennia, cheek by jowl with other mediated modes of human expression and interaction; its origins, in a current reading of surviving evidence, are best understood as a representation neither of speech or of voice; indeed, it is not even clear that we should regard them as linguistic at all; in the beginning, writing was not "an extension of speech" but "an extension of drawing."[6]

But numerous efforts to imagine him (or her: in Rudyard Kipling's "How the First Letter Was Written," the inventor is a little girl) are revealing all the same, for like most mythical points of origin, he embodies tensions that endure in the tradition we would trace back to him.[7] This book will approach that tradition at a point after the invention of writing[8]—and in particular, of alphabetic writing—has fully been accomplished and writers have begun to take the most basic tools of their trade for granted. Nevertheless, we shall find in some of the texts they produced an enduring doubt about writing's root purpose. To be clear, that doubt has not entirely left us, even today, though Edison's wizardry has distracted us somewhat from it. Antiquity's remoteness from us, even more than its relative proximity to writing's origins, will help to bring this doubt to the fore in the pages that follow, even in the case of more recent comparanda.

The tension essential to this doubt may perhaps best be understood in terms of the alphabet itself, the single elements of which are capable of expressing sounds that are less than words, but which, as an ensemble, simultaneously makes possible the inscription of something that is *more* than (mere) language. This latter category is dominated by that class of texts we have come to call "literature," after the Latin word for the very letters (*litterae*) of the alphabet. In such texts, literary heights plunge back to their alphabetic base in search of such "sound effects" as alliteration: to give us more than words, the writer calls our attention to what is less than one. Far older than writing itself, "alliteration was one of a number of phonetic figures available to the Indo-European poet,"[9] or to put this slightly more carefully, the poet's deliberately dense repetition of consonants long predates the invention of the letters that represent these consonants in writing and so give "alliteration" its name. Any writing system that corresponds to fairly stable conventions of speech can capture such phonetic figures (i.e., one hears them when reading), but alphabetic writing actually represents them (i.e., they are as visible as they are audible, at least to the extent to which its letters continue to match, one-to-one, the constitutive sounds of speech). This has led one scholar to call the Greek alphabet, capable of representing both consonants (aspirated and unaspirated) and vowels, "the first technology capable of preserving by mechanical means a facsimile of the human voice."[10] The same scholar goes on to argue that the Greek alphabet must have been devised for the express purpose of writing down poetry, perhaps that of Homer.[11]

The thesis of a sudden poetic origin for the Greek alphabet has found little favor among those who argue instead for its gradual emergence around more prosaic tasks—and in any case, recent evidence would seem to leave little doubt that the Greek alphabet predates the transcription of the Homeric poems themselves.[12] The question of the alphabet's origins, though, may not really be the most interesting part of this puzzle. Sooner or later, *someone* began to write down, for example, the first lines of the *Odyssey*: "Tell me, Muse, of that wily world-traveler who so often was driven off course, once he had sacked the sacred city of Troy." And what we resolutely cannot know is whether that writer sought *primarily* to capture the address to the Muse that survives in my translation, or, instead, the music of the poem's repeated consonants and metrical vowels, lost in translation but able, respectively, within limits we may momentarily ignore, to be transliterated into our own alphabet and transcribed into the relative durational values of modern musical notes:

> Ἄνδρα μοι ἔννεπε, Μοῦσα, πολύτροπον ὃς μάλα πολλὰ
> πλάγχθη ἐπεὶ Τροίης ἱερὸν πτολίεθρον ἔπερσε...
>
> 𝅗𝅥 ♩ ♩ | 𝅗𝅥 ♩ ♩ | 𝅗𝅥 ♩ ♩ | 𝅗𝅥 ♩ ♩ | 𝅗𝅥 ♩ ♩ | 𝅗𝅥 x
> *andra moi ennepe mousa po- lutropon hos mala polla*
>
> 𝅗𝅥 ♩ ♩ | 𝅗𝅥 𝅗𝅥 | 𝅗𝅥 ♩ ♩ | 𝅗𝅥 ♩ ♩ | 𝅗𝅥 ♩ ♩ | 𝅗𝅥 x
> *plankthē e- pei Troi- ēs hie- ron ptoli- ethron e- perse*[13]

One way or another, writing recorded not only the singer's sense, but also no small part of his sensuous sound, both of which continue to drive the poem forward long after its opening lines, even while pushing its protagonist and namesake extravagantly off course.

Naturally, we could say much the same about the continuing odyssey of classical poetry in Homer's wake. Indeed, looking back over the combined poetic-alphabetic tradition, it seems hard to imagine any point (including its point of origin) at which letters seemed the building blocks of language alone.

In one sense, this is a new book about a very old subject: the role of sound in Greek and Latin literature. Indeed, we soon shall find ourselves in the well-charted territory of classical meter, rhetorical figures, and other sound effects to which antiquity gave both form and nomenclature. The stakes of our inquiry, however, will be rather different from those of the various manuals long familiar to students and scholars in classics. In laying claim to these and other disparate sounds as symptoms of textual vocality, I shall be arguing for them a role that is anything but ornamental, that is, a role hardly reducible to that of anodyne "sound effects" or even to that of poetic "musicality," the two explanations generally invoked by said manuals. In addition, I shall follow the lead of medievalist Paul Zumthor in simultaneously distinguishing this *vocalité*, "the corporeal aspect of texts" and so "their mode of existence as objects of sensory perception," from *oralité*, which he defines as the voice's instrumentalization as, instead, the bearer of language.[14] What, however, remains of the voice when it is distinguished both from language and from (mere) sound, and what, if anything, can this remainder tell us?

In search of an answer, we can take some first hints from recent work on the voice by theorists and philosophers, most notably Adriana Cavarero, *A più voci: Filosofia dell'espressione vocale* (2003; trans. 2005 as *For More than One Voice: Toward a Philosophy of Vocal Expression*), and Mladen Dolar, *A Voice and Nothing More* (2006). As their titles already suggest, both introduce the problem of the voice as one of measure. For Cavarero, who finds her initial inspiration in a story by Italo Calvino, voices are always multiple, each of them "a unique voice that signifies nothing but itself," that is, "the vital and unrepeatable uniqueness of every human being."[15] This single, specific voice upends the world of the eavesdropping king of Calvino's story, and it would do the same for philosophy, if the latter stopped to listen, which it almost constitutionally cannot do, for "the philosophical tradition does not only ignore the uniqueness of the voice, but it also ignores uniqueness as such, in whatever mode it manifests itself."[16] Against this monolithic (and monophonic) tradition, Cavarero launches a "challenge" that she will partially derive from the role of the human face in the thought of Emmanuel Lévinas. "This challenge ... consists in thinking of the relationship between voice and speech as one of uniqueness that, although it resounds first of all in the voice that is not speech, also continues to resound in the speech to which the human voice is constitutively destined."[17] Although I began my research for this book in the same spirit of challenge and thus attune to the same polyphony, from my ancient material soon emerged competing vocalities that were both collective and irreducible. "Every sound we make is a bit of autobiography," notes classicist and poet Anne Carson, but as her own title, "The Gender of Sound," suggests, sometimes the sounds we make tell stories that some of us (at least) share.[18] As we shall see, ancient efforts to write *a* voice were almost always also efforts to write *the* voice, conceptualized in varying degrees of generality. Not all of these efforts unfold as preludes to an inevitably disembodied, metaphysical leap to speech-as-*logos*. Cavarero's appeal to listen for the unique voice still echoes throughout this book, but it often is answered by voices that, without embracing logocentrism, nevertheless insist that uniqueness is not the only thing that brings human bodies face to face, listening to and for one another. For the most part, the present book aims to vindicate a more general version of Cavarero's challenge to philosophy, more like what Jean-Luc Nancy epigrammatically expresses in the distinction he draws in French between *entendre* ("to hear," but also "to understand") and *écouter* ("to listen"): "Isn't the philosopher someone who always hears (and who hears everything), but who cannot listen... ?"[19]

Dolar instead takes most of his cues from Jacques Lacan, and it may be useful to ask at once whether the same will be true of this book. After all, the Lacanian reader will already have heard something familiar in the parameters I have begun to use: the voice as an object of desire, seducing us with the siren-song of something beyond language, that which Lacan calls the "real." Beginning in the mid-1960s, explicitly shifting his early focus on the "symbolic" and "imaginary" orders to that of the real, Lacan offered a series of overtures to the voice. In one of these, he traces his contribution back a decade, when he had taken up the subject of Daniel Paul Schreber, whose *Memoirs of My Nervous Illness* had famously been analyzed by Freud in "The Schreber Case":

> When I think that in the phenomenology of psychosis, we are still at the stage of questioning ourselves about the sensorial texture of the voice, when simply with the six or eight pages of a prelude that I gave in my article on "A Question Preliminary to Any Possible Treatment of Psychosis," I designated the quite precise approach from which there can be in our day, at the point that we are at, there can be questioned the phenomenon of the voice.[20]

But neither in the earlier article (which had been redacted from the seminar of 1955–56) nor in his later musings does Lacan move much beyond what we would expect from him, namely, an insistence that the voice, like the real, is an effect of the symbolic.[21] Schreber "hears voices" because of what he *knows* (albeit unconsciously), and the same is true of those more ordinary voices we attribute to our senses and the noisy world beyond. In other words, for Lacan, we do not make meaning out of voices but, rather, voices out of meaning, even when, acoustically speaking, there is nothing to hear.

Fully embracing Lacan's view, Dolar describes the pervasive error it is meant to illuminate:

> Bringing the voice from the background to the forefront entails a reversal, or a structural illusion: the voice appears to be the locus of true expression, the place where what cannot be said can nevertheless be conveyed. The voice is endowed with profundity: by not meaning anything, it appears to mean more than mere words, it becomes the bearer of some unfathomable originary meaning which, supposedly, got lost with language.... It should be stated clearly: it is only through

> language, via language, by the symbolic, that there is voice, and music exists only for a speaking being.... The voice as the bearer of a deeper sense, of some profound message, is a structural illusion, the core of a fantasy that the singing voice might cure the wound inflicted by culture, restore the loss that we suffered by the assumption of the symbolic order. This deceptive promise disavows the fact that voice owes its fascination to this wound, and that its allegedly miraculous force stems from its being situated in this gap. If the psychoanalytic name of this gap is castration, then we can remember that Freud's theory of fetishism is based precisely on the fetish materializing the disavowal of castration.[22]

On its own terms, Dolar's logic is impeccable; what instead points to trouble is, let us say, his own first-person voice, thinly masked as the third: "It should be stated clearly..." For what can we call this scolding about "illusion" and "fantasy" if not precisely a "no of the father" (*non du père*), that is, the prohibition that for Lacan inaugurates our entry into the symbolic order, with Dolar himself in the castrating paternal role? Dolar even finishes things off by invoking the ever-authoritative Dr. Freud—can he not have been thinking of Lacan's play on the "name of the father" (*nom du père*)?—here "remembered" precisely in the name of the re-membering fetish. In other words, as Dolar reading Lacan and even I reading Dolar reveal (and one can repeat this game endlessly), the terms of psychoanalysis are never quite so useful as they are when used to describe psychoanalysis itself.

With his own roots not only in Freud but also in the clinical observation of severe psychosis, Lacan himself, even as he turned to the real, never quite shed a view of the symbolic as a necessary prison beyond which lay the madhouse and, farther off and even worse, meaninglessness. But while paranoiacs, the subject of Lacan's 1932 doctoral dissertation, and other psychotics may hear voices, the fact that those voices so often tell them what to do—stab a famous actress, in the case study at the center of that dissertation—makes them exactly unlike the extralinguistic voice conjured by Dolar and dismissed as the actual "illusion." In the end, what psychotics and psychoanalysts have in common is a ready willingness to assume that truly meaningful voices are always *telling* them something. What if, sometimes, this belief is instead the illusion, one fostered precisely by hermeneutic enterprises like that founded by Freud?[23] In other words, what if it is not meaning that requires language, but only interpretation? The exclusion of such a distinction is foundational to psychoanalysis: even when words are not forthcoming, as in the case of the unconscious and its manifestation in dreams, some kind of language, or at least something like one, is always already there, even when it manifests itself in the form of "a primitive language without a grammar," as Freud puts it in his *New Introductory Lectures on Psychoanalysis*.[24] Lacan less hesitantly pronounces that "it is the whole structure of language that psychoanalytic experience discovers in the unconscious."[25] Interpretation, in such a view, remains close to its root Latin sense of "translation," from one (kind of) language (or one kind-of language) to another. But should psychoanalysis have the last word here?

Psychoanalysis is not, of course, the only practice to reach us from the past century with a deep stake in interpretability, that is, the linguistic translatability of meaning. Let us limit ourselves to the way in which the latest Lacanian word on the "voice and nothing more"—namely, that it is little more than an afterthought of language itself—resembles an independently persistent axiom of twentieth-century literary criticism. From the latter's long bookshelf, let us select just one well-worn but now dusty text: William Empson's *Seven Types of Ambiguity*, which first appeared in 1930 and which continued to be printed in a series of editions into the 1950s. Empson pointedly professed little interest in Freud (whose *Civilization and Its Discontents* was first published in the same year), but something else bothered him far more:

> And if one is forced to take sides, as a matter of mere personal venom, I must confess I find the crudity and latent fallacy of a psychologist discussing verses that he does not enjoy less disagreeable than the blurred

> and tasteless refusal to make statements of an aesthete who conceives himself to be only interested in Taste.[26]

Mostly forgotten today, Empson's book nevertheless helped to train generations of readers and critics to be suspicious of lovers of uninterpretable beauty. Empson begins by taking aim at a particular breed of hermeneutically disinclined aesthete: the kind devoted to what he calls "Pure Sound." Empson aligns his own contrasting position with a celebrated dictum of Alexander Pope: "The official, and correct, view, I take it, is that 'the sound must be an echo to the sense.'"[27] This echoing, however, will not always obey clear principles, Empson notes; in the end, even "a sound effect must be interpreted"; when it cannot be, it should not hold our attention for long.[28] Interpretation, for Empson, is the rational response to all manner of poetic ambiguity and thus to poetry tout court; fans of "Pure Sound" unnaturally resist this hermeneutic drive. But *why* do they do so? Empson compares their irrationality to synesthetic disorders "due to migraines or epilepsy or drugs like mescal" (i.e., peyote), which in turn may offer the dim recollection of "an infantile state":

> Mescal-eaters have just that impression common among readers of "pure" poetry, that they are seeing very delightful but quite new colours, or knowing something which would be very important and interesting if they could make out just what it was. But how such a disturbance can be of serious importance to a reader of poetry is not easy to see, or how one is to be sure when it is occurring.[29]

What fails here is not meaning—or, at least, not (apparent) meaningfulness—but, rather, interpretation, first in the purple haze of the mescal-eater who cannot say "just what it was" (like the aesthete who cannot "make statements"), and then, with the briefest hint of longing, in the reader-critic's inability "to be sure," though proleptically he already has pulled himself together and moved on to matters of "serious importance." Dolar, as we already have seen, will summon the "structural illusion" of purely vocal meaning through a similar picture of earnest error: "The voice is endowed with profundity: by *not meaning anything*, it *appears to mean* more than mere words, it becomes the bearer of some *unfathomable originary meaning* which, *supposedly*, got lost with language. . . ." Unlike Empson, Dolar finds this illusion endlessly interesting, but that it is nothing but an illusion is something on which they agree. And in exposing this illusion, both are led to conjure the benighted true believer who, encountering something that feels like meaning but which cannot be interpreted, wrongly supposes that it must mean something pretty deep, man.

What if the true believer is right? Poets, at least, have long been kinder to addicts and other hard-core aesthetes—and not only because they are an important part of their fan base. "How sweet," writes Tennyson in "The Lotos-Eaters," ventriloquizing the drugged deserters of Homer's *Odyssey*,

> To hear the dewy echoes calling
> From cave to cave thro' the thick-twined vine—
> To watch the emerald-colour'd water falling
> Thro' many a wov'n acanthus-wreath divine!
> Only to hear and see the far-off sparkling brine,
> Only to hear were sweet, stretch'd out beneath the pine.

Of course, on a first listen, these lines might seem abundantly to satisfy any demand that sound echo sense, since the sailors' words mime those "dewy echoes" that enchant them: "calling . . . falling," "cave to cave," "thro' the thick," "twined vine . . . divine . . . brine . . . pine," "Only to hear . . . Only to hear."[30] But the paradox is that sound thus *is* the immediate sense of their words, and their only deeper meaning is this: stop worrying about what it all means, or better, stop worrying about what we were meant to be doing, where we were meant to be going. Among the many things that are happening here is a rejection of the tortured teleologies of epic: as the sailors abandon their odyssey ("we will not wander more" is the last thing they and the poem say), so Tennyson borrows his setting from Homer but his pastoral scene and its dense repetitions from the *Idylls* of Theocritus, which famously begin with a sweetly sonorous pine.[31]

But if the history of literary taste can be boiled down to a long debate about form and content, embracing endless subdebates about (this) form versus (that) form, (this) content versus (that) content, and of course, form itself versus content itself, then the eaters of mescal and lotus, as readers, pose a very particular challenge to the critic. Not merely preferring extralinguistic sound *to* meaning but embracing it *as* meaning (Dolar's "structural illusion" of the voice), they raise the unsettling suggestions that it is instead interpretability that is a "structural illusion" of the literary text, that literature, as a medium, is only incidentally concerned with the recording of language per se, and that literature may best be regarded the use of language itself as a medium, for the recording of something not linguistic at all.

Before his death in 1892, Tennyson enthusiastically allowed himself to be recorded by the phonograph, reciting "The Charge of the Light Brigade," soon followed by other poems.[32] Earlier, one of the first Europeans to write about Edison's new machine, the inventor W. H. Preece, had already appropriated a line from the poet's "Break, Break, Break," lamenting the death of a friend, in order to observe that now at last it would be possible to enjoy forever "the sound of a voice that is still."[33] In his enthusiasm Preece rather misses the poignant ambiguity of "still" (let us finally give Empson his due!), and indeed, that of the entire seaside poem, which both does and does not capture the sound of the crashing waves, "the thoughts that arise in me," the shout of the boy "for his sister at play," the song of the lad "in his boat on the bay"—all proxies for the poet's groans of grief for the still, but still echoing, voice of his friend.[34] When Tennyson wrote these lines, Edison had yet to be born; for Friedrich Kittler what made the latter's invention revolutionary was its ability to mediate just the sorts of things the poet before him could not fully grasp, the extralinguistic real. But the fact that the early phonograph repeatedly went looking for Tennyson hints that the new device was at least partially recognized as a reinvention of "the dark round of the dripping wheel" of his kind of poetry.[35] One can of course object that poets have never really been able to capture the sound of a crashing wave, a sailor's song, or a dead friend's voice—or even their own. But is this ambivalence of reference the same as a failure to record? After all, if the real is beyond the symbolic, then surely its mediation need not be confined by symbolic categories.

Tennyson represents something of an acoustical high-water mark in English literature, but in a moment we shall be throwing open the floodgates of classical literature. Everyone knows that the latter, which calls its poems "songs" (a matter to which we shall return), is a noisy business, offering a battery of sonic devices that are deployed with a density and pervasiveness rarely matched, even when imitated, by later literatures. As we have already noted, however, outside rather vague gestures at "orality" or "musicality," few ever bother to ask why. Already on the basis of our brief considerations thus far, however, it seems possible to hazard a new guess. If, as Dolar suggests, what we call voice is the real that swirls around speech, then the sounds of poetry—not in spite of their lack of clear referentiality, but because of it—would seem to add up to voice, or at least to a partial claim to represent such. And this means that the sonority of classical poetry (and prose, as we shall see) contributes to an especially strong version of such a claim, indeed, one that would long remain paradigmatic, through to the classicizing Tennyson himself. This strong vocal claim is the "ancient phonograph" of my title.

To be clear, the sounds of ancient literature, though a crucial part of this claim, were not themselves sufficient to complete it. In antiquity, the text's phonographic status was supported, for example, by the development and adaptation of oral practices (like oratory itself, subject of Chapters 1 and 5, as well as music, subject of Chapter 3) in ways that maximized their perceived susceptibility to writing, and also by education that controlled for consistency among readers. Other factors are there in the texts themselves, from the stories they tell about voices (like that of Echo, subject of Chapter 2) to the marked tendency in some genres (like tragedy, subject of Chapter 4) to push beyond the speakable but not the voiceable. Such factors became mutually reinforcing, across genres, periods, and the Greek and Latin languages—boundaries likewise crossed by ancient meters, rhythms, and sonic figures.

It is not, of course, that other literatures do not make vocal claims, even strong ones, as Zumthor reveals, for example, for medieval poetry. Still, the sheer extent and diversity of antiquity's phonographic literature is hard to parallel. Naturally, that literature's coherence as "a" literature has been constructed retrospectively, though such retrospection was well under way in antiquity itself. Accordingly, this book will end with two looks back, one from late antiquity, the other from the early Renaissance. As we shall see (and hear), antiquity's specific ensemble of phonographic technologies would, for later ages, be key to its audibility as "ancient" (and, to use the value-laden term, as "classical"). So, too, in antiquity itself, had those same technologies made literature audible as literature. On this last score, scholars sometimes advise, *caveat lector*: antiquity had no consistent term for "literature," and our own use of the word in its most common current meaning is itself not much older than Tennyson and his age. Indeed, for one scholar, Florence Dupont, the application of "literature" to antiquity is to be reckoned among the modern "crimes" against ancient "orality"—that is, against a culture in which texts, in her reading, were either quietly ancillary to performance or performance's mostly mute afterthought.[36] Surely, though, a more serious crime is Dupont's, banishing voices from the very texts in which ancient readers so plainly heard them. (Evidence that they did so will be abundant in the coming chapters.) Against Dupont, this book not only finds voices in classical literature but also finds voice at the heart of literature's classical definition. The classical literary text emerged, in antiquity, not in spite of voices, nor even for the voice's sake, but *as* voice, written.

Why, however, does the voice *matter* to, in, and as (classical) literature? The word I have just italicized, ostensibly to emphasize its meaning but also to conjure the voice in one of the few, clumsy ways favored by academic prose, offers a key: literary voice restores matter to written language, or rather, it calls our attention to matter that, in one form or another, was there all along. To be sure, it is not that texts do not have their own native materialities; indeed, the present book to some extent offers a follow-up to my study of these in *The Matter of the Page*. Voice, however, pointedly presents itself as, or as something like, the matter of the human body, or at least of part of it. (Which part will be a subject of Chapter 1.) To what end? Each chapter will offer a slightly different answer to this, but all will pursue an intuition that what draws us to vocal media is not just *what* they mediate but *that* they mediate. For the living voice, I shall argue, is itself a medium; like the wax of Edison's later cylinders, or that of an ancient writing tablet, its ability to express depends in part on its ability to be impressed. In other words, voice and record are phonographically linked not so much as original and copy as they are by a more basic resemblance rooted in the stuff of which they are made. In this regard, phonographic inquiry goes looking for a certain strand of ancient materialist thought and, from time to time, uncovers a version of what James I. Porter has famously dubbed the "material sublime." Indeed, I record here a blanket acknowledgment of indebtedness to Porter's work, in which the voice has long been a major theme, most recently in *The Origins of Aesthetic Thought in Ancient Greece: Matter, Sensation, and Experience* (2010); he has influenced what follows well beyond my specific invocations of him. That said, not everything to come will be lofty or even entirely literary. From babbling babies, to nursery rhymes recited by inventors, to something like a nursery rhyme half-sung by God himself, this book's soundtrack mixes the sublime with a decent dose of the ridiculous. In the end, my aim is to demonstrate the ordinariness of the voice in classical literature, where it is as much a part of the furniture as are wax and papyrus. This is not to say, however, that writing the voice *well* was not extraordinarily hard work, as we shall see.

One final note on my approach. Maurizio Bettini, in *Voci: Antropologia sonora del mondo antico*, reconstructs antiquity's "phonosphere" by placing human voices and languages firmly against a broader backdrop of animal sounds, with special emphasis on attempts by the former to mimic and transcribe the latter.[37] So too Mark Payne, in *The Animal Part: Human and Other Animals in the Poetic Imagination*,[38] repeatedly invites us to think—and listen—our way outside the category of the human in ancient literature and life. Together they may be allowed to strike a salutary note of caution about what follows. The literary soundscapes that await us in the coming chapters offer no shortage of noisy non-humans, from gods and demigods to birds and beasts to resounding rocks and bubbling brooks. One might well ask, therefore, why we should we seek in these the *vox humana* and not, at the same time and among other things, the call of the wild. For the semantically superfluous sounds of literary texts would seem, *prima facie*, to record that nonhuman call as surely as they do the human one, completing the ancient phonograph's analogy to its modern counterpart, promiscuously receptive to the entire sounding world. One might go farther still, observing that the human voice cannot really assert its distinctiveness without simultaneously revealing its continuity with that same world, which just as surely assails our ears and which can—and, indeed, should—command our attention and care.

Nevertheless, this very homophony between voice and world, precisely because it contests the supposed distinctiveness of human beings from their environment, is what enables a text to construct a human voice out of seemingly nonhuman sources of sound. And what causes a text's varied acoustical elements to coalesce, foremost, as human voice is the simple fact that even (and especially) the most enchanted literary *selva oscura* is a world made not of woods but of words (even if not entirely reducible to these *as* words). As Susan Stewart puts it, "It is not just sound that we hear; it is the sound of an individual person speaking sounds.... Such sounds might be imitations of sounds in nature, of animal cries, or of the most elaborately inflected nuances of human conversation, but in every case sound is here known as a *voice*."[39] This is not to say, however, that this same voice, as it emerges, cannot also renew our sense of a broader, less human-centered ecology. If this book remains focused instead on the humanness of literary texts, it does so out of a low rather than high humanism, one that seeks in literature not boisterous confirmation of human exceptionalism but a quieter invitation to read, and to live, more humanely.

But enough preliminaries. Time now to start taking some (very) old records off the shelf. Our playlist will be an eclectic one, and I mostly have permitted myself simply to choose a few of my own personal favorites: top-forty hits from Cicero, Ovid, and the ancient tragedians, but also some rarer material from sophistic declaimers, pseudo-Anacreon, and even the half-mad emperor Nero. There is some method to my own madness, however, for these selections will permit us to cover a range of phonographic genres: oratory, dialogue, epic, lyric, drama. Although this is not a book about recovering ancient music, a complicated matter best left to experts, some of whom have been producing some remarkable results of late, these last two categories (lyric, drama) will take us some distance into the world of ancient song. By the end, I hope to have reconstructed a basic but fully resonant working model of the ancient phonograph of my title, available for others to spin different sets. Along the way, however, I shall try not to lose sight of my own responsibility to the listening pleasure of my reader. And so: drop the needle, play . . .

TRACK ONE

Body and Soul

Let us begin what will be a fairly noisy book in the (relative) quiet of a library. We are in Athens in the late second century CE, in a Greece already for a long time part of Rome, now ruled by Commodus, son of Marcus Aurelius. Here we find Julius Pollux, professor of rhetoric, who has stolen time from a busy teaching schedule in order to return to his labor of love, the *Onomasticon*, a vast dictionary organized not alphabetically but thematically. "This is so that any weariness which the reader may feel from what he has read may be overcome by his desire to hear what comes next," he explains in a dedicatory letter to the emperor, his former pupil, at the start of the work's ninth book, loosely concerned with vocabulary for the spaces and mechanisms of social interaction.[1] It is on this book, in fact, that we now find him working, and though he soon will turn to lighter subjects such as games, of which he will provide a long list, complete with basic rules, he is at present in the midst of a long chapter that opens, "It may not prove unhelpful to say a thing or two about money."[2] Who can argue with that?

Though at first you might not guess it from his subject matter, Professor Pollux, in his day, was something of a star. Breaking with the rigidly Classical and Athenian canon of his predecessors, his scholarship opened vistas onto the language and literature of the wider Greek-speaking world, both past and present.[3] Back in the classroom, he seems to have attracted devoted throngs of students. Alas, such talents also won Pollux bitter enemies, starting with Phrinicus, the conservative rival in preference to whom he had been appointed to the prestigious imperial chair of rhetoric at Athens. Pollux has long been assumed to have provided, wholly or in part, the model for Lucian's contemporary satire *A Professor of Public Speaking*,[4] which pretends to urge a young student to shun any uphill climb to Lady Rhetoric and to prefer instead a gentle, winding, flower-lined path:

> If you turn to the other road, you will find many people, and among them a wholly clever and wholly handsome gentleman with a mincing gait, a thin neck, a languishing eye, and a honeyed voice, who distils perfume, scratches his head with the tip of his finger, and carefully dresses his hair, which is scanty now, but curly and raven-black . . .[5]

This is the unnamed professor of the title, who, at the end of the preposterous lesson that ensues, will explain that, though he once was called "Pretty-boy," he now "shares a name with the sons of Zeus and Leda," that is, Castor and *Pollux*.[6]

Lucian's portrait is all too predictably imbued with satiric convention to tell us much, but at least one detail, the professor's "honeyed voice" (*melichron phōnēma*), though something of a topos—compare Homer's Nestor, "from whose tongue flowed speech sweeter than honey"[7]—may still be drawn from life, for Lucius Flavius Philostratus, in his account of Pollux in his *Lives of the Sophists*, uses almost the same phrase.[8] Philostratus never heard Pollux, but he evidently got an earful about him from his own teachers in Athens and elsewhere, some of whom were old enough for their careers to have overlapped with that of the controversial professor. Philostratus begins his account by trying to strike a compromise: perhaps Pollux both did and did not know what he was talking about. He spoke, Philostratus continues, with more "daring" (*tolmē*) than "art" (*technē*), leaning too heavily on the "natural talent" (*phusis*) that, one nevertheless had to admit, he had in abundance. "Natural talent" presumably includes some things that Philostratus could not know firsthand, but it clearly also designates qualities that he assumes we can all assess from written records of his speeches.

In fact, he next gives us two excerpts from these, introduced as specimens of a style that was mostly mediocre but here and there redeemed by "trickles of pleasure" (*hēdonōn libades*). The first, barely a snippet, is from what Philostratus calls a *dialogos*, probably a popular type of didactic but entertaining "lecture"[9]:

> Proteus of Pharos, that marvel in Homer, puts on many and manifold shapes, for he rises up into water, blazes into fire, rages into a lion, makes a rush into a boar, crawls into a serpent, springs into a panther, and when he turns into a tree, grows leaves for hair.[10]

The second quotation is longer and comes from a "declamation" (*meletēma*), a practice with origins in classroom exercises but which had evolved into a major public art form: a speaker or series of speakers would deliver theoretically extemporaneous speeches on a given argument (*hupothesis*), sometimes chosen on the spot by the audience. Pollux had been asked to address an imagined assembly of impoverished islanders contemplating a desperate plan to sell their children into slavery in order to pay their taxes. He closed his speech by asking the islanders to pretend to read with him a future letter home from one of their enslaved children. It is this "letter" that Philostratus quotes:

> I am a king's slave; I was given to him as a present from a satrap; yet I never mount a horse of the Medes or handle a Persian bow, nay I never even go forth to war or the chase like a man, but I sit in the women's quarters and wait on the king's concubines. Nor does the king resent this, for I am a eunuch. And I win their favour by describing to them the seas of Greece, and telling them tales of all the fine things that the Greeks do; how they hold the festivals at Elis, how oracles are given at Delphi, and which is the altar of Pity at Athens. But pray, father, write back to me and say when the Lacedaemonians celebrate the Hyacinthia and the Corinthians the Isthmian games; when are the Pythian games held at Delphi, and whether the Athenians are winning their naval battles. Farewell, and greet my brother for me, if he has not yet been sold.[11]

Almost as if surprised and seduced by his own examples, Philostratus follows them with this plea for open minds and ears: "Neutral listeners (*adekastōs akroōmenoi*) can inspect the quality of this man's products. By 'neutral' I mean listeners lacking prejudice in either direction." He closes with the report regarding Pollux's delivery: "He is said to have delivered such things in a honeyed voice (*melikhra phōnē*) with which he even charmed the emperor Commodus, obtaining from him the chair at Athens."

The *Lives of the Sophists* functions as a kind of jukebox, or better still, radio program, with the narrator himself as deejay, interspersing a variety of greatest hits and other singles with anecdotes and commentary in his own voice. As one scholar has noted in this regard, Philostratus thus mimics his material, offering what reads as a recorded oral performance about recorded oral performances.[12] The whole show evokes for us several key features of classical antiquity. Vocal performance was of central importance; this was hardly confined to music or theater; rhetoric, in fact, was in many ways the most important of antiquity's performative arts; this, however, cannot be reduced to oratory's political or juridical functions; in particular, the "sophistic" oratory that Philostratus spins seems best understood under the modern rubric of "entertainment." Though he again and again returns to a frame that is biographical, historical, and critical, Philostratus cannot stop inviting us, implicitly or even, as here, explicitly, to sit back for a moment and slip into the groove. (Indeed, Philostratus closes his introduction with the hope that the effect of what follows will be like that of the care-soothing "Egyptian drugs" Helen famously mixes into her guests' wine in the *Odyssey*.)[13] In precisely this regard, however, he would seem to walk a fine line between what he—and, therefore, we—can and cannot hear. In other words, his project presupposes that orators can be enjoyed (or not) and evaluated on the basis of written records of their speeches, something that would seem to contradict in part the advice of the great Greek orator Demosthenes, who when asked to name the three most important things in oratory, famously replied that they were delivery, delivery, and delivery. Philostratus cannot "play" this for us, and even his descriptions thereof, like that of Pollux's "honeyed voice," are usually secondhand.[14]

A remarkable echo of this dilemma is there in the second "single" selected by Philostratus, which extracts for transcription precisely that part of the speech which purports to be a reading of a text, one that, as a letter, itself stands in for direct speech made impossible by absence. On the one hand, we have in this example a chicken-and-egg question familiar to Derridean ears: Which comes first here, writing or speech? On the other hand, we have a similarly complex chain of thrown voices: Philostratus quotes Pollux, who is dead and whom he never heard in person; Pollux in turn ventriloquizes a boy who is a puppet of his own making. In the end, what is most striking here is that none of these deferrals and displacements shake Philostratus's confidence that we can *hear Pollux* in this example; indeed, as he makes plain, the only real obstacle to this would be a failure on our part to listen objectively.

To understand the reasons for this confidence, let us briefly turn the dial. As antiquity fades away, a more recent oldie crackles into tune: Bobby Vinton singing "Blue Velvet." No one who knows his rendition of the song can read that title or the song's lyrics without hearing the voice that, strictly speaking, words alone cannot record. But were one to need to describe Bobby Vinton's voice to someone who has never heard his most famous recording, what phrase could be better suited to the task than precisely this one: "blue velvet"? Art here thrives on a perceived harmony between form and content; examples of the same effect can be multiplied almost infinitely, throughout the history not only of the vocal arts but of art generally.

Back on Radio Antiquity, we were listening to something similar. Deliberately or not, Philostratus invites us to suppose that the "trickles of pleasure" he urges us, as readers, to enjoy were drizzled from the same honeypot that once coated Pollux's living larynx.[15] He does not tell us whether these are to be found in the passage's saccharine (indeed, schmaltzy) play on sentiment or in the sweetness of its language—but this is just the blurring of another line between form and content. Indeed, if we understand form in its root sense to be the "shape" that an artist gives to (subject) matter, then in the manufactured worlds of ancient declamation, content, outside the basic premise handed to the orator, is itself a kind of form. Needless to say, such shapes did not always align, and even Pollux may often have declaimed sweetly on subjects either savory or sour; at the same time, in indicating that Pollux delivered "this kind" of speech "in a honeyed voice," Philostratus implies that his voice was not an inflexible given but, rather, that its flavor or at least its intensity could be varied to suit the occasion. But what principally led to the success of this particular declamation must have been its multiple layers of sweetness—laid on so thick, here and elsewhere, as to nauseate Pollux's salty-tongued critics. Indeed, in the end, we cannot really know which layer most pleased the emperor's palate, since the alleged criteria relayed by Philostratus sound suspiciously like an attempt to discredit a reasonable appointment, made on the basis of *technē* that Pollux could teach, as, instead, an impulsive imperial prize for *phusis* that he could not. Whatever qualities of Pollux's voice were innate, in other words, might have attracted groupies but would have been of minimal use to serious students. Among the latter, of course, he had once counted Commodus himself.

To some extent, these convergences and confusions can be traced down to a root ambiguity in the Greek word *phōnē*, the principal meanings of which include not only "voice" but also the human faculty of "speech."[16] It is not that the Greeks failed to observe a distinction between these. That Philostratus means the former here, for example, is clear from context: "He is said to have delivered such things in a honeyed voice (*phōnē*)." In other words, this is the *phōnē* that Philostratus has not heard, cannot read, and thus cannot (re)-record. But in the word's other sense, Pollux's "speech" is exactly what Philostratus gives us, at least to the extent that a written text can register this. The result, I would suggest, is a dual positioning of the former, unwriteable *phōnē*. On the one hand, Philostratus presents it here as an almost insignificant afterthought. But on closer inspection, we might be inclined instead to regard it teleologically, as a vanishing point of our shared desire to hear. In other words, a particular speaker's *phōnē*, in its full sense, is something that Philostratus both can and cannot capture for us. This makes it very like the "lives" of the mostly dead sophists in his title. After all, one cannot really make the absent present through "biography," that is, by writing (*graphein*) life (*bios*) down. Neither can this particular miracle be achieved by newer vocal media like the "phonograph" (*phōnē* + *graphein*)—though one of the most startling early claims for Edison's invention was that it would console us in our grief with "the familiar voices of the dead."[17]

Our post-Edisonian age has come to regard and even define the voice as something *outside* speech (and a fortiori outside writing): either before it (we have a voice before we learn to speak), below it (voice is the material support of spoken language), or beyond it (the voice expresses what language alone cannot). Most definitions, in fact, assume all three and imagine the voice as something that surrounds speech without exactly entering it. *Phōnē*, however, takes us back to a moment, or at least a language, in which this exile (or is it a siege?) had yet to receive its final declaration. We can dig deeper still, to the word's hypothesized Indo-European root, **bhā-* (or **b^{h}eh$_2$-*), "to speak," source also of the regular Greek verb for speaking, *phēmi*. So too, linguists tell us, was **wek^w-* the root of the Latin *vox* (and thus of our own *voice*) and meant "to speak."[18] Far from being outside speech, the voice seems almost to emerge from within it. Of course, we should hardly be surprised to encounter some deep mystery in the making of words for the making of language. But that mystery is maybe not so mysterious after all. Long before the invention of writing, what cause would there have been to distinguish "speech" from the rest of our utterances, especially since, then as now, these too could be meaningful? Perhaps we can reach back farther still, much farther, in fact, to an almost unthinkably distant age in which the most important thing our voices did was to identify us to one another as members of the same species. Eons later, whether crying like a baby or crooning "Blue Velvet," our call still always "says" this: I too am human.

What of texts and other talking machines? "I'm sorry, Dave, I'm afraid I can't do that": 2001 has come and gone, but we still seem resigned to a future in which these will rebel against us, making our voice fully their own, usurping our humanity, controlling our lives (or ending them). Perhaps, however, this rebellion came long ago, already in the first spring of human writing. Consider the ancient reader, who, though hardly unable to read silently (despite the stubborn persistence of this scholarly myth), was rather more likely to read aloud or to be read to, including by trained readers, than we are.[19] Such a relationship between vocalizing reader and mute text would seem to leave little doubt as to who lends what a voice, and even the breath of life. But on closer inspection, who—or what—is playing whom? The confidence of readers, ancient and modern alike, that reading is a social act ultimately depends on the assumption that living, present readers animate the words of dead or otherwise absent writers. The text, in this sense, is a moderately miraculous medium between two worlds, but the transaction, finally, is between two human beings. So believed Pollux, as he gathered *testimonia* for his *Onomasticon*; so too Philostratus, as he compiled his *Lives*. But there in the library—I, too, speak from experience—one cannot help wondering: What if it is not the dead who move my lips? What if it is just the books themselves? As Jesper Svenbro brilliantly observes, silent reading actually *increases* the written text's claim to *have* a voice, since understanding that voice does not require the help of any *other* voice—except that of the "interior voice" in which you yourself have just been hearing my (?) italics.[20] Even now, are you really listening to me, or is it just the bibliomachine talking?

Better leave the library before we become paranoid, though arguably this last vision is no madder than supposing that the voices we hear, as we read, are those of people who are not there. My point, however, is simply to suggest that the understanding of texts as *media*—again, in its root sense as something in the "middle" between ourselves and some other human being—forever defers the question of whether texts are themselves human. This repressed possibility corresponds to longstanding fears, which of course are also longstanding desires. In fact, in these dreams that can turn into nightmares survives an older—and, I would suggest, truer—understanding of the human voice as, quite simply, the sounds made by a human, the sounds we recognize *as* human. Of course, in the literary wood, like the wilderness of our real-world wanderings, we may not always be sure of that voice's source: Is another flesh-and-blood person calling to us, or is this a voice that somehow emerges in and out of the text itself as a human presence never attached to any real human body? A real forest that emits voices on its own is, like treacherous supercomputers and machinating libraries, potential cause for alarm: Wild beasts are imitating our call! Ladies of the lake are luring us to our doom! (Alternatively, as Lucretius rather devastatingly suggests, we may simply be trying to convince ourselves that our world has not been deserted by the gods.)[21] Nevertheless, while there can also be something uncanny when human sounds seem to come from the leaves of books, there is a quieter sense in which the voice can be key to a text's ability to attract and sustain our attention at all, even for the most untroubled work of reading.

But, gentle reader, before we return to our literary playlist, we must turn our ears toward the sometimes ungentle voice of philosophy. We find the twentieth-century author of *Of Grammatology* as we found the second-century author of the *Onomasticon*, viz., writing while he reads:

> All the metaphysical determinations of truth, and even the one beyond metaphysical ontotheology that Heidegger reminds us of, are more or less immediately inseparable from the instance of the logos, or of a reason thought within the lineage of the logos, in whatever sense it is understood: in the pre-Socratic or the philosophical sense, in the sense of God's infinite understanding or in the anthropological sense. Within this logos, the original and essential link to the *phonè* has never been broken.[22]

Soon thereafter, Derrida introduces what will become Exhibit A(ristotle) of his case against the "phonocentrism" and "logocentrism" (sometimes combined as "phono-logocentrism") he has just described:[23]

> ἔστι μὲν οὖν τὰ ἐν τῇ φωνῇ τῶν ἐν τῇ ψυχῇ παθημάτων σύμβολα, καὶ τὰ γραφόμενα τῶν ἐν τῇ φωνῇ.

This is the second sentence of Aristotle's influential treatise *On Interpretation*, which Derrida quotes in the French translation of Jules Tricot:

> Les sons émis par la voix sont les symboles des états de l'âme et les mots écrits les symboles des mots émis par la voix.[24]

So, too, does Derrida's own English translator, Gayatri Spivak, borrow the sentence from a long-standard translation, that of E. M. Edghill:

> Spoken words are the symbols of mental experience and written words are the symbols of spoken words.[25]

There is an interesting and persistent problem of Derridean ambiguity mirrored by the two versions, one which we can summarize by noting that the *voix* of the title of his *La voix et le phénomène* (1967, like *De la grammatologie*) is rendered as "speech" in the work's first English translation (1973) but as "voice" in a more recent one (2011). Nevertheless, the switch in the quotation from "émis par la voix" to "spoken" properly depends not on Derrida's ambiguity (though it is true that he does not really distinguish "voice" from "speech" in either work), but on Aristotle's, which itself is a consequence of something now familiar to us, namely, the dual meaning of *phōnē*.

There are, in fact, a number of complexities in Aristotle's extraordinarily dense sentence, of which we shall require a more careful translation than that offered by Tricot or Edghill.[26] The best way in is perhaps through the three prepositional phrases. The first and third of these are *en tēi phōnēi*, which, as we have seen, Tricot translates as "émis par la voix" and Edghill as, simply, "spoken." Between them is *en tēi psukhēi*, where the object is another complicated Greek word, *psukhē*, "soul," though it sometimes corresponds to what we might be more inclined to call "mind," for which, however, Aristotle also has recourse to a narrower term, *nous*. (We shall return in a moment to the place of this in Aristotle's tripartite theory of the soul.) *En tēi psukhēi* modifies *pathēmata*, from the class of Greek words that describe "passivity" itself: "things suffered" or "felt"; collectively, "experience." Our two translators combine the phrase with the noun to produce, respectively, "états de l'âme" or "mental experiences." Their translations thus undo the sentence's insistent parallelism: *en tēi phōnēi* . . . *en tēi psukhēi* . . . *en tēi phōnēi*.

In fact, it is this middle phrase that is the key to unlocking the image lurking behind the whole sentence: that of the *psukhē* as something like wax, pressed upon by the sensory world. This popular simile is taken up by Socrates in the *Theaetetus* of Aristotle's own teacher Plato,[27] and Aristotle himself employs it in his brief treatise *On Memory*:

> One might ask how it is possible that though the affection (*pathos*) is present, and the fact absent, the latter—that which is not present—is remembered. It is clear that we must conceive that which is generated through sense-perception in the soul (*en tēi psukhēi*), and in the part of the body which is its seat,—viz. that affection the state whereof we call memory—to be some such thing as a picture. The process of movement stamps in, as it were, a sort of impression of the percept, just as persons do who make an impression with a seal.[28]

These *pathē en tēi psukhēi*, in other words, are imagined as impressions in a soul imagined as an impressionable material. So, too, are the almost identically named *pathēmata en tēi psukhēi* of *On Interpretation* impressions. They thus anticipate the *graphomena* ("things written," but presumably modifying an understood *sumbola*) at the sentence's very end, for antiquity's paradigmatic writing material was the waxed tablet, itself often used, in a variant of the simile we have just seen, as a figure for the soul and memory. In the end, this materialism embraces *phōnē* too, which here hardly designates incorporeal *logos* but, rather, the stuff of speech, the body of the voice.[29] This, finally, is the function of the thrice repeated preposition *en*, not really captured by the "par" of Tricot or the participle of Edghill: the entire sentence is about the media (*phōnē*, *psukhē*, and the tablets implied by *graphomena*) "in and on" which impressions can be made.

All of this looks very much like grist for Derrida's mill, revealing the passage to be nothing less than an Aristotelian precursor to Freud's "Mystic Writing Pad" (*Wunderblock*), the toy in which one writes with a stylus on a thin sheet of cellophane that is pressed into a tablet of black wax below, used by Freud as a metaphor for the "psychic apparatus." In "Freud and the Scene of Writing" (first published in 1966 and republished in *Writing and Difference*, the other volume of the 1967 trilogy), Derrida seizes on the "Mystic Writing Pad" as a glaring example of "the metaphor of writing which haunts European discourse," promising instead to offer this:

> An attempt to justify a theoretical reticence to utilize Freudian concepts, otherwise than in quotation marks: all these concepts, without exception, belong to the history of metaphysics, that is, to the system of logocentric repression which was organized to exclude or to lower (to put outside or below), the body of the written trace as a didactic and technical metaphor, as servile matter or excrement.[30]

In the case of Aristotle's sentence, however, that repression and exclusion have been even more emphatic in translation and analysis, including Derrida's own. Still, Aristotle's use of the participle *graphomena*, rather than a fourth prepositional phrase (e.g., *en tēi graphēi*) that would have made the parallelism complete, arguably already sets in motion the subsequent deracination of his psychic figure from its root metaphor. This would seem to vindicate the instincts that led Derrida here, even via Tricot's misleading translation.

A problem lingers, however, for we have not yet exhausted the metaphors of Aristotle's sentence. To what he finds "in the voice," Aristotle applies the term *sumbola*, which Tricot and Edghill (and Derrida and Spivak) translate by transliterating as "symbols." Long before it came to mean "symbol," *sumbolon* meant instead this:

> tally, i.e., each of two halves or corresponding pieces of an ἀστράγαλος ["knuckle-bone"] or other object, which two ξένοι ["guest-friends"], or any two contracting parties, broke between them, each party keeping one piece, in order to have proof of the identity of the presenter of the other.[31]

As others have observed, Aristotle never leaves this meaning far behind. Specifically regarding *On Interpretation*, it has been suggested that he compares words to these authenticating "tallies" to invoke "the sphere of convention and social agreement" in which, a few sentences later, he will position language itself.[32] Such a level of abstraction, however, seems oddly dissonant with the waxen meditations that characterize the rest of the sentence. What would it mean instead to make Aristotle's metaphor depend on *sumbolon* in the fully material sense of which he and his age had not yet lost sight?

It was this question that first led me to the chapter on money in book 9 of the *Onomasticon* of Julius Pollux, who there explains that "half of a coin split in two" (*hēmitomon ti nomismatos*) could function as a *sumbolon*.[33] If we suppose that Aristotle had this apparently common class of *sumbola* in mind, then we might be tempted to read his metaphor as comparing a monetary economy to a verbal one—except that making a coin into two *sumbola* takes it out of general circulation, which would somewhat undo any such metaphor's point.[34] But coins have characteristics beyond their use as currency. Consider what surely is the most famous figural use of *sumbolon* in its root sense in Greek literature, from the speech of Aristophanes in Plato's *Symposium*. When his turn comes to improvise on the dinner party's chosen topic of love, Aristophanes humorously suggests that Zeus once punished the rebellious human race by splitting our original, double-facing, four-legged bodies in two, condemning us as *sumbola* to a life of erotic searching for our corresponding halves:

> ἕκαστος οὖν ἡμῶν ἐστιν ἀνθρώπου σύμβολον, ἅτε τετμημένος ὥσπερ αἱ ψῆτται, ἐξ ἑνὸς δύο· ζητεῖ δὴ ἀεὶ τὸ αὑτοῦ ἕκαστος σύμβολον.
>
> Each of us is a *sumbolon* of a human being, split into two halves, just like flatfish. And we are forever looking for the *sumbolon* that matches us.[35]

Taking *sumbolon* in the basic sense of "tally" gets the general point, but a finer one is had by supposing that Plato's Aristophanes is riffing specifically on *sumbola* made from portrait-bearing coins, offering almost dizzying play on the question of where the middles of both coins and human bodies lie: rejoining two halves of a coin produces a complete portrait, except that this is usually in profile and thus offers only half a face, which even when doubled and attached to the rest of its body turns out to be (in this tall tale) just half of an original being that faced opposite directions simultaneously, like the obverse and reverse "faces" of a coin seen instead in the round. One way or another, the story zeroes in on an irreducible human middle that cuts all the way down the fronts of our torsos (toward which our faces, originally facing back, are turned at the end of this primeval surgery), a matter to which we shall return in a moment.

To suppose that Aristotle too had numismatic *sumbola* in mind would of course add another kind of mark-making to his brief sentence's ensemble of writerly imagery—further pleasing Derridean readers. This, however, yields a dilemma. The inscriptions on a coin are produced prior to its division into *sumbola* and reappear only after its reunification. Inscription, in other words, is secondary to the two main events in a *sumbolon*'s life; it may help to ratify that a join is exact, but it is ultimately optional, as indeed is confirmed by the fact that uninscribed objects (like knucklebones) were used as *sumbola* too. Properly speaking, therefore, if we are to take the Aristotelian *sumbolon* as an emblem of writing, we must look first to its broken edge. Indeed, the peaks and valleys there suggestively resemble those made by a die in a coin blank, or better, by a seal or stylus in wax, even as their mirroring across the break conjures just the sort of perfect symbolic correspondence that Derrida claims Aristotle is here foisting, via a writerly fantasy, onto speech. In another sense, however, this zigzag is exactly unlike writing: random, unique, irreproducible. Its specific contours are the result not of the metallic strength of an instrument that has been pressed into it but of its own intrinsic and hitherto invisible material weaknesses. In fact, as I now hope to show, a rush to see writing here—or even speech—would miss something far more important.

A more delicate sense of the join between *phōnē* and *psukhē* in Aristotelian thought can be had from his treatise *On the Soul*, which breaks with Platonic precedent and presents the soul as an inescapably embodied thing:

> Hence the rightness of the view that the soul cannot be without a body, while it cannot *be* a body; it is not a body but something relative to a body. That is why it is *in* a body (*en sōmati*), and a body of a definite kind.[36]

This is a classic example of Aristotelian "hylomorphism," his signature view (un-Platonic and, we might almost say, deconstructive) that "matter" (*hulē*) and "form" (*morphē*) cannot exist apart and should not be theorized separately:

> That is why we can dismiss as unnecessary the question whether the soul and the body are one: it is as though we were to ask whether the wax and its shape are one, or generally the matter of a thing and that of which it is the matter.[37]

Here again is the familiar metaphor of wax, though its application has shifted: no longer like wax *in which* impressions are made, the soul is compared to those impressions themselves. One way or another (the slippage in metaphors is, in a sense, the hylomorphic principle in action), the soul is always in a body (*en sōmati*). Later in the same treatise, Aristotle will explore what that body shares with the one in which we find the voice:

> Voice (*phōnē*) is a kind of sound characteristic of what has soul in it; nothing that is without soul utters voice, it being only by a metaphor that we speak of the voice of the flute or the lyre....
>
> The organ of respiration is the windpipe, and the organ to which this is related as means to end is the lungs. The latter is the part of the body by which the temperature of land animals is raised above that of all others. But what primarily requires the air drawn in by respiration is not only this but the region surrounding the heart. This is why when animals breathe the air must penetrate inwards.
>
> Voice then is the impact of the inbreathed air against the "windpipe," and the agent that produces the impact is the soul resident in

> these parts of the body. Not every sound, as we said, made by an animal is voice (even with the tongue we may merely make a sound which is not voice, or without the tongue as in coughing); what produces the impact must have soul in it and must be accompanied by an act of imagination, for voice is a sound *with a meaning*, and is not *merely* the result of any impact of the breath as in coughing . . .[38]

We would seem here to be at the heart of Aristotle's "phono-logocentric" problem, with the spoken and the psychic bound together at the postulated core of meaning-making. But *phōnē* here emphatically is not (merely) "speech," or even exclusively human. Indeed, as similar passage in his *History of Animals* observes, not all of spoken language belongs to the voice:

> Voice and sound are different from one another; and language (*dialektos*) differs from voice and sound. The fact is that no animal can give utterance to voice except by the action of the pharynx, and consequently such animals as are devoid of lung have no voice; and language is the articulation of vocal sounds by the instrumentality of the tongue. Thus, the voice and larynx can emit vowel sounds; consonantal sounds are made by the tongue and the lips; and out of these language is composed.[39]

This passage makes it even easier to see that Aristotle's primary point in both is neither hermeneutical nor metaphysical but anatomical, pushing the voice down from the mouth and tongue and up from the diaphragm and lungs, into the "windpipe" (*artēria*, including the pharynx and larynx, somewhat hazily placed). The voice is generated "by the soul resident in these parts of the body," and this phrase is key. Aristotle located the "sensitive soul" (*psukhē aisthētikē*) of blooded beings in a particular part of the body, namely, the heart.[40] The heart, in turn, he placed "above the lung at the division of the windpipe."[41] Like the lungs, "the heart also is attached to the windpipe, by connexions of fat, gristle, and sinew; and at the point of juncture there is a hollow. When the windpipe is charged with air, the entrance of the air into the heart, though imperceptible in some animals, is perceptible enough in the larger ones."[42] Combining all these passages, we can now see that it is this contact and even interpenetration (Aristotle's anatomical errors only emphasize what he is looking for) that make the voice possible, embedding at least the base of the windpipe in ensouled cardiac flesh.

Aristotle famously assigns three kinds or aspects to the soul: the nutritive, the sensitive, and the rational or "noetic" (from *nous*, "mind"). The voice has no need for proximity to the first of these, since the heart sends nourishing blood throughout the body. That the last too does not require a soul-voice connection is clear from the fact that many irrational beings (namely, animals with just the first two kinds of soul) have, for Aristotle, voices. It is the middle, sensitive soul, therefore, that acts in the windpipe to produce a voice from air. This soul is the seat of "feeling," in the full range of the English word, from simple sensation to emotions, *pathēmata* to *pathē* (though Greek, as we already have seen, did not always keep these two neatly distinct). It produces a voice, Aristotle tells us, through its access to *phantasia*, "imagination," which in Aristotelian terms is the flip side of *aisthēsis*, "perception": the latter is a product of stimulus; the former works without such.[43] The sensitive soul, capable of receiving impressions, is therefore also a sensible soul, capable of making impressions, that is, capable of making itself felt, in and as the voice. So, too, is the voice both sensitive (receiving impressions from the soul) and sensible (making impressions of its own). While we experience the voices of others almost entirely though our ears, we *feel* (haptically) our own voices in our throats and chests. Consider the singing Sirens in figures 2 and 3, perhaps sculpted in Aristotle's own lifetime.[44] How do we know they are singing? Not merely because their mouths are parted, but because *they* know they are singing: pointing to, embracing, enveloping the voices they feel in their own terracotta bodies.[45] Whatever else he was doing in all the passages we have reviewed, Aristotle was looking for this voice: the sensitive-sensible voice rising up from the sensitive-sensible soul, all crammed into a span of human flesh not much bigger than a fully opened hand.

Figures 2 and 3. Sirens (details), from Sculptural Group of a Seated Poet and Sirens, Terracotta with Polychromy, South Italian Greek (Tarentine), 350–300 BCE, Getty Villa, Malibu, 76.AD.11. Images courtesy of the J. Paul Getty Museum.

Let us return, as promised, to love. When we say "I love you" with a pounding heart, or "Don't leave" with our heart in our throats, or "I miss you" with a heart that aches or has broken, we feel the words we say more or less in the same places where we feel the longing, fear, or pain that inspires them. "That's the pain, / cuts a straight line / down through the heart; / we called it love," as Hedwig paraphrases Plato's Aristophanes on "The Origin of Love" (see Discography). Of course, any conviction that what is in our soul (or hearts, as we would say) and what is in our voice are the same thing may well add up to the "metaphysics of presence" from which Derrida sought to cure Western thought. The simpler belief that these are in the same part of our body, however, offers both less and more: a phenomenology of the embodied voice not without roots in—dare we say it?—physiological facts. For we really do feel some very important emotions in our (racing) hearts and (quickening) breath and (tightening) throats, and we really do hear their effects not only in our own voices but also in those of others. Deep down, this is why Aristotle compares what is "in the voice" and what is "in the soul" to two halves of the same fractured matter: sometimes our hearts and our voices seem to break in precisely the same place. These are, of course, exceptional events, but they point us to the broader truth that speech is not simply something we use but, rather, something that happens to us, *in* us, in the Promethean clay of our bodies, whether that resonance is in our ears or in our deeper middles, generated by other souls or our own. Voice is the name we give to that event.

This, let us say, is the Aristotle who opens *On Interpretation*—or, at least, this is where a more productive reading of that treatise might itself begin.[46] While translation cannot find easy equivalents for that opening's delicate metaphors of wax and metal, knucklebones and clay, there is at least one matter we share: that of the body. Of course, the suggestion that we share a body not just with the Ancient Greeks but with the physiologist who gave us wet women, dry men, and other catastrophes of Western thought may give us pause. Still, we can surely find common ground in this particular pound of pounding flesh, paraphrasing Aristotle's point of departure in these more familiar terms: there is a place—(in) the body—where what we feel and what we say meet.[47] But what of their match? Though soul and voice are of one flesh, Aristotle is emphatic that their *sumbola* have no original, natural unity. Each pair is instead imposed "by convention" (*kata sunthēkēn*), that is, extrinsically, by the specific language that it is our lot to learn:

> I say "by convention" because no name is a name naturally but only when it has become a tally (*sumbolon*). Even inarticulate noises (of beasts, for instance) do indeed reveal something, yet none of them is a name.[48]

The translation of *kata sunthēkēn* as "by convention" is itself a matter of convention, obscuring Aristotle's clever wordplay. For *sunthēkē*, "convention" or "compact," is derived from a verb that means "to put together" (*suntithēmi*) and thus describes quite literally what one must do with matching *sumbola*, which, in the word's extended sense, can likewise designate "contracts" and the like. In other words, Aristotle is saying that we come together to put together words with things; we agree to make this agree with that. Behind these single matches is therefore an aggregate one that parallels that of the voice itself: if the recognition of a human voice tells us one thing about the origins of a sound, the recognition of Greek speech tells us another.

In various ways, we might say, a *phōnē* will tell us where it comes from: this soul, this soil. At the same time, confidence in something we think we share may lead us to say to one another, "I know where you are coming from," even when we find ourselves at the limits of language. Those limits can be human ones: that thing for which we have not yet agreed upon a word, or the sobbing that does not even look for one. Or they can be the limits of humanness itself: the "inarticulate noises" of wild animals, which nonetheless "reveal something." At the root of our ability to understand each other at all is, for Aristotle, a fundamental belief that we live in the same world. Here, he is anything but a constructivist: "But what these are in the first place signs of—affections (*pathēmata*) of the soul—are the same for all; and what these affections are likenesses of—actual things—are also the same."[49] Our shared world yields shared experiences and even shared suffering: in a word (though not one used here by Aristotle), "sympathy" (*sumpatheia*), without which our *sunthēkē* of *sumbola* would not be possible.

After all this, why does Aristotle go on to produce, in *On Interpretation*, such a disembodied account of human language and logic? Part of the reason lies in his own truest sympathies, a clue to which is to be found in the "inarticulate noises" (*agrammatoi psophoi*) of wild animals. *Agrammatoi* here means, foremost, "unwriteable," but it also offers a playful personification: these are beasts that have not learned their (Greek) letters, animal analogues to human barbarians and illiterate Greeks alike. In *On Interpretation*, Aristotle is not speaking of or with any of these. Indeed, though his subject is human language generally, he simultaneously is crafting, here and throughout his *oeuvre*, an idiolect for a chosen few. About the language of "the Philosopher," which would become the language of philosophy itself, let us say just this: there is little that is "honeyed" about its voice. *Pathos mathos*, as the Greek proverb goes, "to suffer (is) to learn,"[50] and for many centuries we have learned philosophy by suffering through Aristotle's noetic prose—and that of his successors, through to Derrida.

A more important reason lies in the hierarchy implicit in Aristotle's title subject. The object of Aristotelian "interpretation" is *logos*, which in the treatise means something like "a complete sentence," standing metonymically for the realization of connected discourse. The treatise's often-quoted (but frequently misrepresented) definition—"*logos* is *phōnē* that signifies" (λόγος δέ ἐστι φωνὴ σημαντική)[51]—is itself a remarkably condensed example of just such a sentence, offering in essence a doubly hylomorphic proposition that again depends on the double meaning of *phōnē*. On the one hand, a sentence is the form that finally emerges from a composite of the parts of "speech": single syllables and the "nouns" (*onomata*) and "verbs" (*rhēmata*) they form. But in *On the Soul* Aristotle had already proposed that "*phōnē* is sound (*psophos*) that signifies"[52]—that is, meaning begins to take form already in still deeper matter of the "voice." Aristotle thus presents *logos* as the sum and summit of all this meaning-making, the final form of all the stuff of which language is composed. This really means that *logos* is just a back-formation from Aristotelian interpretation itself: *logos* is whatever can be interpreted in the terms here presented. Thus, the involuntary cough that Aristotle ostentatiously suppresses (twice) in *On the Soul* is anything but *logos*. The physician might be able to "interpret" it, but not the philosopher.

However, it is another rejected human sound that will help us to diagnose Aristotle's true malady: "With the tongue we may merely make a sound which is not voice."[53] This mindless clicking seems a strange variant on the unvoiced consonants (*aphōna*) that, as we have seen, Aristotle likewise locates solely in the tongue and lips, like the ones that lie on the very *t*ips of our *t*ongues: *t-t-t*, *th-th-th*.[54] What, we may ask, finally distinguishes these from the alliterations of, say, the great orator Gorgias, wondering why we are surprised (*ti thaumaston?*) if Helen of Troy succumbed to the irresistibly divine power of love (*theos . . . theōn theian*)?[55] So, too, do the rejected growls of beasts seem to be a distant and distracting substitute for, say, the melodic groans of a tragic chorus. In other words, more interesting than the sounds Aristotle explicitly rejects for interpretation are ones like these that he never mentions. For they suggest that the principal threat to Aristotelian *logos* comes not from farther down the slopes of sound but, rather, from the rival peaks of Parnassus or Helicon: loudly indifferent to *agrammatoi psophoi*, Aristotle's *On Interpretation* also but more quietly turns its back on the sounds of art. These, of course, do receive airtime in Aristotle's twin works of criticism, the *Rhetoric* and *Poetics* (though hardly their due: Gorgias will be panned for producing poetic nonsense rather than oratory, and music will be ranked fifth in importance among the six essential elements of tragedy).[56] But this only confirms that they lie outside the bounds of philosophy proper.

Constitutionally deaf, since Aristotle, to sounds that do not themselves constitute meaning, interpretation has often proven itself to be a remarkably unfeeling lover of the Muses. *Ti thaumaston*? After all, "each man kills the thing he loves," to borrow a line from that latter-day Gorgias, Oscar Wilde.[57] Indeed, there arguably is something of a goddess/whore complex (to say nothing of a Platonic one) in Aristotle's representation of interpretation as the serial rejection of single words (*logoi*) in order to reveal the *logos* that they are not.

There is an alternative to this cold-bloodedness, in the different kind of "phono-logocentrism" too briefly invoked by Aristotle himself at the start of *On Interpretation*—and never imagined by Derrida. Let us call this "philology" in its once-upon-a-time sense of a "love of words" both promiscuous and heartfelt, a tireless romance with the body (and bodies) of language. For this lover, every sweet sound is a potential source of pleasure (*Sì, mi chiamano* ***Mimì***, sings the consumptive heroine of Puccini's *La Bohème*), while even a cough may tell him something he does not want to know. The mode (and mood) of the present book will continue to be, in this sense, a philological one.

Nevertheless, we shall borrow much from the philosopher, starting with his hylomorphism. We shall apply this to the twin quantities embedded in the Greek word that, consciously or unconsciously, may well have inspired the principle in the first place—namely, *phōnē*. In other words, we shall take as axiomatic that there is no language without matter. In the coming chapters, we shall exercise this axiom largely on written texts, and to the extent that we can say that these "speak," we shall also say that they have "voices": *phōnē*-graphs in the full sense of the word. This last assertion may at first surprise modern ears, long used to phonographic rivals to the text that were immediately celebrated for their ability to capture all voices and all the voice:

> This little instrument records the utterance of the human voice, and like a faithless confidante repeats every secret confided to it whenever requested to do so. It will talk, sing, whistle, cough, sneeze, or perform any other acoustic feat. With charming impartiality it will express itself in the divine strains of a lyric goddess, or use the startling vernacular of a street Arab.[58]

Strikingly, however, this early review begins and ends with the voice used for speech (rather than song, which would prove to be the phonograph's truest calling), just as Edison's first patent for his device was for an "Improvement in Phonograph or Speaking Machines."[59] "Improvements" is a legal dodge (others were working on similar devices, and the Frenchman Charles Cros had planned but not realized something very close a few months earlier[60]), but the emphasis on speech reveals that Edison and his contemporaries saw the phonograph not as the *first* technology to record and play the voice, but simply as a much *better* one than ordinary writing.

Itself unlettered, Edison's phonograph easily captures *agrammatoi psophoi*, including those in and around the voice; even the elusive cough does not escape its stylus. As we later shall see, texts travel with and in contexts of use that can themselves be remarkably sensitive recorders; still, "unlettered" sounds would seem to lie definitionally outside the bounds of the alphabetic text itself.[61] We shall first turn, therefore, to sounds that very much can be written down. What will distinguish these from records of speech as language? Here too we shall take a cue from Aristotle, but we shall reverse its direction, seeking what comes not at but below the level of *logos*: single words, syllables, letters. Where meaning begins to fade, we shall suppose, the voice as the matter of speech may become more audible. In other words, we shall listen for something rather like "the startling vernacular of a street Arab," that is, sounds that we cannot (yet) interpret. What we shall hear—in literary Greek and Latin—might startle the philosopher but not the (true) philologist: namely, abundant evidence that linguistic meaning is hardly the only end of speech. At the same time, we shall also listen for things beyond speech, like "the divine strains of a lyric goddess." In this regard, the written text confronted a different rival in musical notation, known in antiquity but persuasively developed only in the Middle Ages. But as with the phonograph, this represented a relative (if revolutionary) change rather than an absolute one, for ancient writers had a range of means other than notation for representing music and song, as we soon shall remind ourselves.

In antiquity, phonographic texts and live vocal performance (musical and otherwise) enjoyed a symbiotic relationship that went far beyond the question of original and copy, score or record, a matter to which we shall return. But it may at once be objected that there is a difference between "recording" a voice and crafting one *tabula rasa*, and that the latter is the rule in (written) literature, works of which need not correspond to any prior act of speech or song. In reply, let us simply note that, variously mixed, corrected, and synthesized, modern records too falsify the apparently singular event they purport to record; only a few deserve to be called "live," and even these have their tricks.[62] Seldom the "faithless confidante" imagined by the phonograph's early reviewer to repeat all it hears, the record's true treachery has forever lain in its omissions, amplifications, and wholesale inventions. However, in any world that takes its media seriously, "artificial" is never the same as "unreal": a mediated voice, true or false, is a voice all the same. With this in mind, we shall approach the ancient voice in its full range, from texts that present themselves as records of actual performances, like the snippets of Pollux rerecorded by Philostratus, to poetic voices explicitly forged in the (real) wax of a writing tablet.

What, however, of the real in the Lacanian sense—that is, that which lies beyond the symbolic? As we have seen in the introduction, "voice" and "speech" are taken by some to map this very tension. Our entry into language is said to banish us from the realm of the real, which, however, still seems to surround us, fleetingly seen, suddenly remembered, taunting us with its proximate distance. For media theorists like Friedrich Kittler, what made the phonograph revolutionary was its promise to capture this siren-song ("the divine strains of a lyric goddess"), recording the real without recourse to the symbolic. By contrast, literary invocations of the real have long been seen by certain Lacanian critics as inescapably frustrated acts of desire: what language cannot do is a fortiori out of reach for literature.[63] We have already begun to trouble this view by suggesting that the text is a vocal artifact that cannot be reduced to language, and more signs in this direction will come into view as we proceed.

We may summarize the basic problem thus: reductive as it may aim to be of everything else, the symbolic can never quite succeed in reducing *itself* merely to its symbolic function; something more always remains. For the interpreter, this residue is mere noise—but this is only because she or he is listening to and for something else. We shall instead call this hermeneutic noise the voice. The voice, let us say, is the real of the symbolic, which, even as it seemingly enshrines and enforces absence and alienation, is itself always present to us. Perhaps, as the Lacanians insist, there is no real presence *in* language, but the presence *of* language is a different matter, and it is this ever-present soul mate, vibrating in our guts and beating with our hearts, at least as much as any endlessly protracted deferral of something we long for but cannot again have, that keeps us forever talking, singing. To say nothing of reading, writing: even Pollux's *Onomasticon* gives us not merely a symbolic vitrine of *sumbola* and other objects we cannot touch, but a treasure trove of words that we can—in our lungs, tongues, and all between.

TRACK TWO

Falling in Love Again

> In some cultural contexts, interpretation is a liberating act. It is a means of revising, of transvaluing, of escaping the dead past. In other cultural contexts, it is reactionary, impertinent, cowardly, stifling. . . . Today is such a time, when the project of interpretation is largely reactionary, stifling.
>
> —SUSAN SONTAG, "AGAINST INTERPRETATION"

> For *we know* every toil on Troy's wide plain
> by Greeks and Trojans when gods so willed,
> all that happens on the nourishing earth, *we know*.
>
> —FROM THE SONG OF THE SIRENS,
> AS REPORTED BY HOMER, *ODYSSEY* 12.189–91

Twice in the twentieth century, the ghost of Narcissus returned from the banks of the River Styx, in whose hellish waters he is said to have kept seeking his beloved reflection after death, in order to haunt important accounts of how a human infant first enters the world of language. The first, with which we will not much be concerned, though it was destined to have by far the greater influence on scholars of art and literature, was that of "The Mirror Stage" of Jacques Lacan, who suggested that, in discovering his reflection, a child won a sense of autonomous selfhood, but at the price of the objectification (of everything, himself included) that is essential to the grammar of the "symbolic" order.[1] The second, famous among linguists but not much cited beyond them, regards instead a remarkable point of resemblance among the languages of the world. The first thing the average Italian child learns to say is *mamma*, that language's "nursery word" for "mother"; the same person is called *mâma* (妈妈) by infant speakers of various forms of Chinese; among indigenous Quechua speakers of the Andes, the word sounds much like English's own *mama*. Not all such resemblances are quite this close, but when hundreds of them were gathered and tabulated in the 1950s by anthropologist George Peter Murdock (who, perhaps not entirely coincidentally, was at the same time secretly gathering and tabulating information for J. Edgar Hoover on possible Communist ties of his colleagues at Yale), they demanded explanation.[2]

What the speakers of the world's many languages share, soon explained the great Russian linguist Roman Jakobson, is parental vanity.[3] The sounds that the world's nursery words (for father as well as mother, and several more besides) have in common are among the easiest for a human infant to make. Sometimes, this has to do with how their mouths are shaped for the far more pressing task of nursing, during which they may be moved to vocalize through their noses; hungry soon again, they may go through the same motions even when food is frustratingly absent. As they continue to explore these and other easy sound-shapes, their loving parents bend down and hear themselves reflected, adapting what they hear to the phonemic patterns of their own languages and repeating it back to their wide-eyed infants, who play along, laughing at a joke which, before Jakobson, we failed to get: what we take to be an infant's first word is actually just our own self-referential echo of his or her still senseless voice.[4]

In the course of illuminating this profoundly important but somewhat embarrassing scene, Jakobson offers a stray observation that sounds so commonsensical that its wisdom is easy to overlook. Without very good reason, Murdock had atomized the nursery words he collected into their single syllables; this obscures, Jakobson observes, the reduplication (*ma-ma*, etc.) that, while not universal to them, is nevertheless strikingly frequent:

> The reduplication of syllables, while passed over in Murdock's test, appears, however, as a favorite device in nursery forms, particularly in parental terms, and in the early word units of infant language. At the transition from babbling to verbal behavior, the reduplication may even serve as a compulsory process, signaling that the uttered sounds do not represent a babble, but a senseful, semantic entity. The patently linguistic essence of such a duplication is quite explicable. In contradistinction to the "wild sounds" of babbling exercises, the phonemes are to be recognizable, distinguishable, identifiable; and in accordance with these requirements, they must be deliberately repeatable. This repetitiveness finds its most concise and succinct expression in, e.g., *papa*. The successive presentations of the same consonantal phonemes, repeatedly supported by the same vowel, improve their intelligibility and contribute to the correctness of message reception.[5]

This "compulsion," in other words, comes from the hovering parents, who, if they are to wrest names for themselves from their child's babble, must first detect repetition. Without Echo, let us say, there can be no Narcissus. Not really speaking at all, the infant is made to produce not just a word, but language itself, in miniature. For *mama* and *dada* are not just baby's first words: they are tiny parental recordings of the sonic minima we require to recognize language as such.

Confirmation that these are the expectations of adult speakers comes from their application to more than just baby talk. Antiquity provides some especially memorable examples. Ancient Greeks called non-Greeks, starting with Persians and Medes, *barbaroi* (whence our "barbarian"), because their languages supposedly sounded like this: *bar bar bar.*[6] The term gives as it takes away, recognizing these foreign sounds as a language but simultaneously infantilizing them as perhaps nothing more than rudimentary repetitions. In Aristophanes, we find instead a group of ghostly frogs who repeat themselves: *brekekekex koax koax.*[7] These strange-sounding creatures, however, are no barbarians, for they also speak Greek; their repetitive croaking can be read, it has recently been argued, as a comic reach upward, through and beyond language, to the iterative sonorities of poetry and music and, farther still, to the ecstatic cries (*Iakkh' ō Iakkhē! Iakkh' ō Iakkhē!*) of Dionysiac cult.[8]

The Romans have instead given us the reduplicated word Jakobson uses to describe the nasal sounds made by a nursing infant, *murmur*, which they sometimes applied to similar scenes. Thus does Hypsipyle, in the *Thebaid* of Statius, mourn the loss of her tiny son and his "incomplete words" and "gurgles (*murmura*) understood only by me."[9] More often, however, adults are the source of the speech at the limits of intelligibility that Latin calls a *murmur*. Let us consider some examples from the *Metamorphoses* of Ovid, a poem characterized by "a kind of phonographic imaginary," as Lynn Enterline puts it in a brilliant study that itself begins with the severed tongue of Philomela, which "murmurs" (*immurmurat*) as it creeps across the floor after her rape and mutilation.[10] Not all murmurs are products of such violence; sometimes, the problem is merely one of volume or proximity. We ourselves, for example, hear (because we read) the doomed princess Myrrha's suicide speech loud and clear, but only "murmurs of words" (*murmura verborum*) penetrate the door to reach her faithful nurse, sleeping outside.[11] Other times, a *murmur* is a jumble of multiple voices that are difficult to distinguish, as when Jupiter silences the one produced by a grumbling assembly of his fellow immortals.[12] Still other times, the word describes the first, futile effort to speak after a bestial metamorphosis, such as that of Atalanta and Hippomenes (transformed into roaring lions) or of Ulysses' men (transformed into grunting pigs), who offer their "murmurs" *pro verbis*, "in place of words."[13]

Ovid tells us that there is no escaping sound in the palace of Rumor, filled not with shouts but with "low-voiced murmurs, like those that come from the waves of the ocean if you are standing far away."[14] He thus links whispered gossip and the like to the natural phenomenon in which he and other Latin writers were most likely to hear a *murmur*: moving water, from the sea, as here, to the "bubbling brook" (*cum murmure labens . . . rivus*) in which the goddess Diana proposes to bathe, to instead the ominously "murmur-less" waters (*aquas sine murmure euntes*) in which the nymph Arethusa attempts the same.[15] Ovid's, of course, is an enchanted landscape, but the use is general: water, thunder, wind, and more all "murmur" in Latin.[16] While such uses invite personifications and other flights of fancy, they also remind us of the opposite, namely, that we are hardly the only things in our world that repeat themselves. Repetition, in other words, may be necessary for us to recognize something as language, but it alone is not sufficient for us to do so.

Everything, however, that we find meaningful (linguistically or not) necessarily has already repeated itself. For the most basic task of cognition is the filtering, within certain limits, of difference, in order to allow the detection of repetition and thus the imposition of identity on things which, under finer measure, are not really the same. The sound waves that we hear as a particular word are never identical, even in repetitions thereof by a single speaker; nevertheless, we hear the same word. Language here necessarily conforms to a more basic rule governing any identifiable sound. Indeed, what is true of sound is true of all our senses. Bigger or smaller than the last, each wave reaching the shore—here I mean "wave" in its ordinary sense—must finally strike us as the same thing we saw (or felt or smelled or tasted or heard) moments before. "You cannot step into the same river twice," declared Heraclitus[17]—but we mostly act as if we can, certain that, at any rate, it is still a river. Plato, of course, ties the world's infinitely diverse stimuli to a barely remembered paradise of pure forms. My interest, however, is not in this or any other fictive unity, but in the (perceived) repetitions that make the imagination of such possible. More specifically, my particular interest is the repetitiveness not of the world, but of art.

We could begin an investigation of repetition in art practically anywhere, starting with the earliest cave paintings, which are full of such. Or we could begin with dance, or architecture, or better still, ornament, where the role of repetition is especially evident. Let us start instead at the top, or at least, what antiquity usually regarded as such: the ethereal summit of Parnassus, mountain of poets. Let us begin specifically in the rarefied air we have just left: that of Ovid's *Metamorphoses*. Here let us go straight for the jugular: the death and dismemberment of Orpheus, legend's first poet and Ovid's most enduring emblem for his own art. Ripped apart by angry Maenads,

> Membra iacent diversa locis; caput, Hebre, lyramque
> excipis, et (mirum!) medio dum labitur amne,
> flebile nescioquid queritur lyra, flebile lingua
> murmurat exanimis, respondent flebile ripae.

> Place divides his limbs; you, River Hebrus, his head and lyre receive;
> and then a miracle: while they float down the middle of the stream,
> something weepy (but what?) is the complaint of the lyre,
> something weepy
> is the murmur of his lifeless tongue, something weepy is the answer
> of the banks.[18]

Never was Ovid's Latin more efficient (my translation requires twice as many words), which only calls more attention to his triple repetition of *flebile*, "weepy," applied first to the sound of Orpheus's lyre, then to that of his voice, and finally to an echo of them both, to which, since *flebile* really means "something that could make you cry," we are implicitly invited to add our own similar sounds. If we place all of this in the hands of an ancient reader inclined to mutter the text as it goes by (at least when the poetry sounds this good), then we may hear yet another murmuring resonance in the *basso continuo* of reading itself. Noisy lines indeed! Yet this sonic multiplication still yields something hard to hear, or hard to understand, or both: a *nescioquid*, that remarkable Latin word for "something or other, but I know not what." Does the lyre play a tune—or just random notes plucked by the current? Is his tongue saying something, and if so, what? Or is its "murmur" just that of the bubbling brook? Are these echoed sounds a song, or is it precisely the absence of song in them that calls for tears of grief? Finally, is Ovid's very silence on these questions designed to make an eloquent point, like that of one of antiquity's most celebrated painters, Timanthes, who in depicting the sacrifice of Iphigenia covered her father's face with a veil in order to suggest his unspeakable (that is, unpaintable) pain?[19]

Walking a line between what he can and cannot hear, can and cannot know, can and cannot say, Ovid leads us to a dizzying Parnassian precipice. But why? For help, we turn to one of the most captivating books ever written about Latin poetry: Jeffrey Wills, *Repetition in Latin Poetry: Figures of Allusion*. Latin poets repeated themselves forwards and backwards, upside down and sideways; they repeated single letters, verbal roots, endings, whole words, phrases, meters, narrative patterns, themes. Wills, however, limits himself to "word-repetition—the close reuse of the same lexical stem," for which choice he gives this simple, elegant reason:

> For our purposes, repetition is a clean, formal feature; it has the advantage of possessing little or no semantic value (unlike diction, where one must consider whether any alternative expression existed, repetition is almost always semantically otiose or at least easily avoided).[20]

For me at least, it is this semantic wastefulness that makes Wills's book a page-turner, like a vast compilation of the greatest hits of poetic sound triumphing over sense. Wills himself, however, is reluctant to enjoy such sound as unsignifying. Rather, his book offers a range of efforts to make sense of repetition, culminating in the argument implied by his subtitle: in many cases, repetition is borrowed from a literary model that is being imitated in other ways too; the conspicuous semantic superfluity of word repetition helps it to underscore the ensemble of allusive connections to which it belongs. Repetition in one sense signals repetition in another.

Wills thus has some intratextual repetition serve the broader needs of intertextual repetition, which scholars since Late Antiquity have tended to treat as an end to itself, offering only marginal improvements when they read it either too lightly as learned "play," or too seriously as a bid for "authority." It is easy to see what is wrong with Wills's thesis (and by extension, the critical tradition just evoked), for no mass of evidence can ever distinguish its validity from that of a far simpler formulation, namely that one kind of repetition simply shades into another, and that both the intratextual and intertextual varieties—in concert or, as is of course more often the case, separately—pursue similar ends. What might these be? In another book, precisely in regard to Orpheus, I have argued that allusion can be read as a macroelaboration of the "backward glance," that is, the retrospection and regression into which the material text inevitably invites us.[21] In the present book, however, our interest is not in looking, but in listening—less Narcissus, more Echo. Let us therefore continue to follow the still enigmatic sounds of our pied-piping poet.

Wills is anything but wrong to suggest that when a text echoes both itself and other texts, the two phenomena may be working in unison. Indeed, this is nowhere clearer than in the Orpheus passage. Brilliantly connecting it to a series of others in Greek and in Latin, including some hypothesized lost verses of Gallus, Wills suggests that they all belong to a miniature tradition of repetition in the midst of full surrender to the so-called pathetic fallacy, when the poet makes all of nature echo human feeling.[22] Ovid's surrender has in fact begun a few lines before:

> **Te** maestae volucres, Orpheu, **te** turba ferarum,
> **te** rigidi silices, **tua** carmina saepe secutae
> fleverunt silvae, positis **te** frondibus arbor
> tonsa comas luxit...

> For **you** the mournful birds weep, Orpheus, for **you** the herded beasts,
> for **you** the stubborn stones, for **you** the woods who often
> followed **your** song, and the tree, its green leaves gone,
> shorn to mourn **you**, grieves...[23]

Compare Virgil, in the *Aeneid*, announcing the death of the priest and soldier Umbro:

> **Te** nemus Angitiae, vitrea **te** Fucinus unda,
> **te** liquidi flevere lacus.

> For **you** Angitia's grove, for **you** Fucinus with its crystal pool,
> for **you** the liquid lakes shed tears...[24]

And earlier still, Theocritus, singing of the death of Daphnis:

> **τῆνον** μὰν θῶες, **τῆνον** λύκοι ὠρύσαντο,
> **τῆνον** χὠκ δρυμοῖο λέων ἔκλαυσε θανόντα.
> **ἄρχετε** βουκολικᾶς, Μοῖσαι φίλαι, **ἄρχετ'** ἀοιδᾶς.
> **πολλαί** οἱ πὰρ ποσσὶ βόες, **πολλοὶ** δέ τε ταῦροι,
> **πολλαὶ** δὲ δαμάλαι καὶ πόρτιες ὠδύραντο.
> **ἄρχετε** βουκολικᾶς, Μοῖσαι φίλαι, **ἄρχετ'** ἀοιδᾶς.

> **Him** the jackals, yes, **him** the wolves bewailed,
> **him** the woodland lion lamented when dead.
> ***Begin****, sweet Muses, your rustic tunes,* ***begin!***
> **Many** the cows at his feet, **many** the bulls,
> **Many** the calves, even newborn, who moaned.
> ***Begin****, sweet Muses, your rustic tunes,* ***begin!***[25]

There is so much here to ponder: the explicitly musical setting with its repeated refrain[26]; the fact that a Greek syllable (*tē*-) crosses, as a different word (though one likewise designating the person lamented), into the Latin sounds of its imitators; and of course the pathetic fallacy to which Wills draws our attention and which places us squarely on the line between repetition as meaning, and language as repetition. Nevertheless, let us forgo any further wandering in this enchanted literary wood in order to concentrate on the one tree in it whose murmuring leaves Ovid necessarily hears above all others: Virgil's description of Orpheus's demise in his fourth *Georgic*. This time we need not content ourselves with a *nescioquid*, for Virgil knows exactly what Orpheus had to say:

> Tum quoque marmorea caput a cervice revulsum
> gurgite cum medio portans Oeagrius Hebrus
> volveret, Eurydicen vox ipsa et frigida lingua,
> a miseram Eurydicen! anima fugiente vocabat:
> Eurydicen toto referebant flumine ripae.

> Even then as his head, torn from his marble neck,
> was rolled along by the Oeagrian Hebrus, carried on the

> gurgling rapids, "Eurydice," cries his very voice and cold
> tongue, "Ah! poor Eurydice!" as his breath of life flees.
> "Eurydice" the banks, along the whole course, echo back.[27]

Eurydice! So too may ancient readers who knew their Virgil have cried out in response to Ovid's imitation, interrupting their reading to offer an impatient correction of the latter's nameless *nescioquid*. Of course, more sophisticated readers, perhaps including Ovid himself, may instead have rebuked Virgil's original, with a question: Really? That is, is something so obvious really the best that we can do? However, let us imagine a third kind of listener, one who knows neither poem and, indeed, has never heard of Orpheus or Eurydice. We can perhaps picture her as a carefree nymph who has come this day to dip her toes in the murmuring waters of the Hebrus. Lo, a severed head floats by! What is it saying? It matters little which version of those sounds reach this innocent bystander, for even Virgil's simple, unmusical "Eurydice" is, for her, a meaningless *nescioquid*. Even if she recognizes it as a name (hardly a given, since it is not a common one), she does not know the girl to whom it once belonged, or her fate. All she hears, to use Virgil's phrase, is the "voice itself," and even he seems to recognize that this *vox* (both "voice" and "word" in Latin) is far less a word than it is a body: a cold tongue (*frigida lingua*) made to resonate by the last sigh of life (*anima fugiente*). Answering that voice with the same physiological response, expressive but not linguistic, with which Ovid instead describes it, our hypothesized nymph weeps.

What, after all, is "in" a name? With this famous question, we must briefly change heroines. Moments before asking it, Juliet has of course cried, "O Romeo, Romeo! Wherefore art thou Romeo?" By this echoed name, she pretends to make her lover present (not yet knowing that he is in fact hiding below her balcony), but for the person thus named, "Romeo" means only one thing, beyond his fondest hopes: *Juliet is speaking to me!*[28] Juliet herself, in other words, is *in* "Romeo," because it is her voice that so invokes him; her word points to her beloved, but her voice is instead indexical of herself, to whom he is called. In other words, "Romeo," to Romeo, here means Juliet, which is perhaps why, bowing to her request to "doff thy name," he suggests, "Call me but love."[29]

Speaking (Shakespeare's) English, Juliet has resorted to the interjection "O" to mark her direct address. By contrast, Greek and Latin have recourse, in such situations, to the vocative case. The vocative is remarkable for being a linguistic form that embodies a kind of extralinguistic performance (to the extent that such an exteriority can be posited). This has to do with more than its name, from the Latin *vocare*, "to call," the verbal cognate of *vox*. For the vocative's preference for a final vowel, though it is little remarked upon, clearly exists to facilitate "calling" and other vocal work in direct address. *Orpheu* is the vocative of *Orpheus*: Where are you, *Orpheuuuuu*? But he can no more hear us than can he himself be heard by the doomed wife to whom he calls *Euridiceeeee*. That his severed head utters her name in the vocative is made clear by Virgil's *figura etymologica*, in which verb echoes subject: *vox . . . vocabat*. But by giving us that call not in direct discourse, but as direct objects (or accusatives of exclamation?), Virgil cleverly damps its sound (*Eurydicen*); as is often the case, he is subtler under close inspection than he first seems.[30]

Ovid instead gives us a vocative, though one addressed neither by or to Orpheus or Eurydice but, rather, by the narrator to the river: "You, Hebrus (*Hebre*), received his head and lyre. . . ." Moments before, however, in a passage we already have seen, the same voice addresses the mutilated poet: "The mournful birds wept (*fleverunt*) for you, Orpheus (*Orpheu*). . . ." That this is yet another direct address that does not really expect to reach its destination only calls more attention to its source, whom we are led to identify with the poet who opens the *Metamorphoses* in vocative address to the gods (*di*), called upon to inspire his poem. The Orpheus story's sudden switch to the past tense (the immediately preceding account has been in the present) points in the same direction, returning us to the historical mode and once-upon-a-time distance with which the whole poem has begun:

> In nova fert animus mutatas dicere formas
> corpora; di, coeptis (nam vos mutastis et illa)
> aspirate meis primaque ab origine mundi
> ad mea perpetuum deducite tempora carmen.

> I am moved to speak of forms changed into new
> bodies; gods, to my beginnings (for these too you changed)
> grant inspiration, and from the creation of the world
> up to my own day, spin me an unbroken song.[31]

Orpheus, too, tells stories about the past (these, indeed, have filled most of the preceding book of Ovid's poem), and it probably goes without saying that when our poet (Ovid) talks to his predecessor (Orpheus, but also Virgil, whom he echoes), he is also talking, poet-to-poet, to himself. The softened voices of both predecessors yield to what is loud and clear: that of Ovid, who must make a name for himself in part by cannibalizing what another poet, Horace, famously calls the *disiecti membra poetae*, that is, the words and phrases of his "dismembered" models.

What's in a name? Whatever is in Ovid's will make it, according to the claim of the poem's closing lines, "uneraseable" (*nomenque erit indelibile nostrum*). Ovid's own name, however, is nowhere in the book that bears it and that is filled instead with a cast of thousands, some of whom lend the story little more than their names. An especially egregious example of this promiscuous naming comes in the story of Actaeon, the doomed hunter who stumbles upon Diana at her bath and is transformed into a deer. Parodying a distinguished tradition of literary lists, like the Homeric catalogue of ships in the *Iliad*,[32] Ovid offers instead a roll call of Actaeon's dogs:

> Dum dubitat, videre canes primique Melampus
> Ichnobatesque sagax latratu signa dedere,
> Cnosius Ichnobates, Spartana gente Melampus.
> inde ruunt alii rapida velocius aura,
> Pamphagos et Dorceus et Oribasos, Arcades omnes,
> Nebrophonosque valens et trux cum Laelape Theron

et pedibus Pterelas et naribus utilis Agre
Hylaeusque fero nuper percussus ab apro
deque lupo concepta Nape pecudesque secuta
Poemenis et natis comitata Harpyia duobus
et substricta gerens Sicyonius ilia Ladon
et Dromas et Canache Sicteque et Tigris et Alce
et niveis Leucon et villis Asbolos atris
praevalidusque Lacon et cursu fortis Aello
et Thoos et Cyprio velox cum fratre Lycisce
et nigram medio frontem distinctus ab albo
Harpalos et Melaneus hirsutaque corpore Lachne
et patre Dictaeo, sed matre Laconide nati
Labros et Argiodus et acutae vocis Hylactor
quosque referre mora est...[33]

While he wavered, the dogs spotted him. First Melampus
and Ichnobates, clever barker, gave signals.
Icnobates was Cretan, Melampus of Spartan blood.
There race the others, faster than the rushing wind,
Pamphagos and Dorceus and Oribasos, Arcadian all,
strong Nebrophonos and fierce Theron, with Laelaps,
Pteralas, good with his feet, Agre, good with his nose,
Hylaeus too, lately wounded by a savage boar,
Nape, born of a wolf, and Poemenis, who once
herded sheep, and Harpy, two pups at her side,
and Sicyonian Ladon, who has girded his loins,
Dromas, Canache, Stricte, Tigris, and Alce,
Leucon with a white coat, Asbolos in black,
Ladon, sturdy and strong, Aello, built to run,
Thoos and fast Lysice with her Cyprian brother,
and Harpalos, who marked his black head with a
spot, Melaneus and Lachne, shaggy neck-down,
and though of Cretan father, a Spartan gave them birth,
Labros and Argiodus, and high-pitched Hylactor,
and those whom it would take long to list...

The reader, ancient or modern, with a little Greek can make ready sense of most of these names: Melampus has a "black foot"; Ichnobates is a good "tracker"; Pamphagos "eats everything"; Dorceus is a "deer-dog"; and so on. In other words, these names (and their accompanying descriptions, which often simply repeat them) tell us exactly what we would expect to hear about hunting dogs; they contribute barely more to our understanding of the plot than does knowing that the third part of Ovid's full name, Publius Ovidius Naso, means "nose" and may therefore be the legacy of a distant ancestor with an especially noteworthy one. Ovid's catalogue is thus like an interruption of the storytelling by so much meaningless barking, beyond the erroneous "signals," *signa*, with which the first two dogs set it off, all part of a deliberate joke by a poet who, like one of that first pair, is a pretty clever barker (*sagax latratu*) himself. The set-up finds its punch line when the narrator breaks off the list, after giving us "only" thirty-five names, with the abrupt suggestion that relating the rest would represent needless "delay" (*mora*). Even amidst this almost slapstick humor, however, it may not be an accident that *mora* is also a theatrical term for the action that occurs when dialogue stops and, even more suggestively, a musical term for a single "beat" of metrical time, especially since this very line gives us five straight dactyls, DA-dada DA-dada DA-dada DA-dada DA-dada, in over-the-top imitation of the hounds' final sprint toward the "prey" that is instead frozen in the line's final spondee, *praedae*:

> quosque referre mora est; ea turba cupidine praedae...
>
> and those whom it would take long to list; that pack with yearning for prey...[34]

For Ovid's comic interlude, I would suggest, invites us to consider what a poem would be if made entirely of proper names we do not recognize,[35] stripped to a *bar-bar* of meter and meaningless syllables, a kind of music without real words. Could Ovid's reader still recognize his master's voice in poetry thus distilled? For at least one very careful reader, William S. Anderson, the slightly less extreme experiment of this canine catalogue provides "an Ovidian tour de force."[36]

Actaeon, of course, loses even his syllables, and Ovid's *mora* has only postponed his now imminent doom:

> Clamare libebat;
> verba animo desunt; resonat latratibus aether.

> He'd like to cry out;
> words fail that wish; sky echoes with barking.[37]

At some point in the course of the poem's transmission, a zealous reader saw here an opportunity to ventriloquize the author and improve the text by inserting between these lines the would-be words of Actaeon's unrealized shout:

> "Actaeon ego sum; dominum cognoscite vestrum!"

> "I'm Actaeon! Recognize your master!"

There are many such interpolations in the *Metamorphoses*, and this one offers some measure of cleverness, since it anticipates the shouted "Actaeon! Actaeon!" of the hunter's companions, soon to search for him in vain, and, perhaps, the *iste ego sum* ("that's me!") of Narcissus, coming soon in the same book. However, in addition to putting words in the mouth of the poet who thought better of giving us the last "Eurydice" of Orpheus, the supplement completely misses the point of the preceding catalogue, since what we shout at our dogs is not our own name, but theirs. Actaeon's problem, in other words, is not the loss of his name but the loss of his voice, which is why his dogs cannot be made to recognize the master they knew but moments before. (Clearly we are meant to contrast Homer's famous depiction of the aged hound Argos who, though Odysseus had been absent from Ithaca for twenty years, "lifted his head and pricked up his ears" in recognition of his own master's voice.[38]) But literary dogs and their owners have a higher master: the poet, whose signature voice prompts both interpolation and efforts to detect such. "This verse, *Actaeon ego sum*," snarls the great seventeenth-century editor Nikolaes Heinsius, who was the first to mark the line as spurious, "is no more Ovid's than it is mine" (*Hic versus,* Actaeon ego sum, *non magis Nasonis est, quam meus*).[39]

In the end, it matters little that Publius Ovidius Naso's own name has avoided erasure, since we know barely more about the man who once answered to it than we do about Actaeon's dog Agre, "good with his nose"—unless we suppose, with the last century's historicist critics, that we can know the poet by studying his prince. Immediately after claiming indelibility for his name, Ovid offers a historicist "reading" of his own:

> ...nomenque erit indelibile nostrum;
> quaque patet domitis Romana potentia terris
> ore legar populi, perque omnia saecula fama
> (si quid habent veri vatum praesagia) vivam.

> ...and my name will be uneraseable;
> and as far as Roman power spreads across earth's vanquished lands,
> I shall be read by the people's mouth, and by this fame,
> across the ages
> (if there be any truth in the predictions of poets) I shall live.[40]

This collective Roman "mouth" has parallels in Latin but strikes our own ears as strange, though it arguably is no less so than language which we instead share with Latin, like the figure of the "people's voice" (*vox populi*), or the idea that many speakers can share the same "tongue" (*lingua*).[41] All of these remind us that a perceived specificity of the embodied voice need not always correspond to a radical specificity of its vocalizing bodies. Anticipating, in jarringly imperialist terms, the more benign Renaissance postulation of a "republic of letters" (*res publica litterarum*), Ovid imagines Roman conquest as the manufacture of media players. The image of these once-barbarian Latin speakers fuels the deliberate ambiguity of "I shall be read," which conflates the rudimentary literacy needed to "read Ovid" in the sense of making out his name (as if the poem were a hulking funerary monument to him) with the more complex skills required to read the whole fifteen-book work.[42] Of course, even the latter may lead us back, in the end, to sounds as elemental as those snatched from the mouths of babes.

Every act of reading a text, Michel Foucault famously argues, inasmuch as it is an act of interpretation, necessarily conjures an "author-function," a hypothesized subjectivity behind the meanings we find there, whether or not these ever crossed the mind of the single person (if there was just one) who wrote what we are reading.[43] Let us likewise suppose, however, a "voice-function," responsible for everything we read that cannot be given entirely over to meaning. When the author-function loosens its hold on the text, the voice-function comes to the fore. Just as the former fashions a mind, the latter fashions a body—out of our own voice, voices we have heard, voices that our reconstructions of the past help us to imagine. These last are dead, as is their language, as is, of course, our author, who ends his poem with *vivam*, "I shall live," precisely because he now hands off the animation of his voice to readers, he hopes, of ages to come. This ending echoes Ovid's beginning, in which he promises to trace, "from the creation of the world up to my own day" (*ab origine mundi ad mea . . . tempora*), the history of "forms changed into new bodies" (*in nova . . . mutatas . . . formas corpora*). Of course, such metamorphosis comes with a risk of silence (think of Actaeon) against which the loud bravado of *vivam* is surely waged. This note of worry in the author-function's last word, however, is only a reflection of what his voice-function has accomplished: a text that has remained far too noisy to ignore. We hear "Ovid" in the poem's every *nescioquid*, elements, let us say, of an unmistakable *je ne sais quoi*.

This, of course, is the kind of recognition any self-respecting poet really wants. The "name" is just a placeholder for the voice that always lurks there—except in Ovid's own name, unless he pronounces it, which he does not, even here, waiting instead for us to do so. This is the same trick by which he sought to make us cry "Eurydice!" To the poem's dying lines we are supposed to answer from our distant banks, "O Ovid, Ovid! Your rows (of verse) by any other name would smell as sweet!" But why call this perfume a "voice" rather than a "style"? The latter, from the Latin *stilus*, "pen," has its own fair claim to be the proper pipe of poetry and, indeed, has been by far the preferred term even for the sonic features to which I have been referring. "Style," however, has suffered from its application to other things that can be done with a pen—namely, drawing and, by extension, the visual arts in general. The result has been a devaluation of its material specificity. A rose, for the voice, is a sound, but in terms of style, it is also an image, a figure. "Rose is a rose is a rose is a rose," writes Gertrude Stein, rejecting metaphor and other semantic wordplay, but delighting, as always, in sound.[44] Style's identification with the semantic work of figuration, that is, with the underscoring of meaning, has tended to place it at the service of interpretation. Far from undermining the reading of literary language as code, style offers itself up for complementary decipherment. Voice, by contrast, is more troubling, as Garrett Stewart suggests in proposing an experiment in what he calls "phonemic reading":

> Stylistics is concerned with the lexicon recruited by syntax, words arrayed in sequence; when it takes up the single letter, it is usually only to calibrate a recurrence or symmetry (such as alliteration) deployed to brace or inflect the syntactic structure. With a quite different importance accorded the letter as signifying marker, phonemic reading enters the perpetual dialectic between lexicon and syntax, their mutual tension and erosion. Stylistics studies the language act; phonemic reading intersects the action of lingual formation itself, its increments and transformations: the former is directed at *parole* (the speech act), the latter at *langue* (the whole structural fund of a given language) in its continuous but shifting provision for utterance. Stylistics is concerned at most with phonological patterns, while phonemic reading takes up their morphological implications, the junctural overlaps and detachments that both make and unmake words.[45]

In a moment, we shall turn to material that seems pointedly to invite just such a reading.

First, let us briefly leave sound behind and return to Narcissus. No literary figure has been more intimately connected with the image than has he. Narcissus is among the favorite subjects of ancient painters, who exploit the occasion to offer an erotic *mise en abyme*: we look at a naked youth looking at his reflection, depicted as a watery abstraction that seems to be dissolving back into paint, a gimmick that would continue to attract painters through Caravaggio and beyond. By the Renaissance, Leon Battista Alberti was teaching his students to regard Narcissus as the founder of the painting itself. Many times since, artists have made their way back to Narcissus in self-reflective encounters either with new media, as in the opening scene of Jean Cocteau's film *The Blood of a Poet*, or with old media they are seeking to transform, as in Salvador Dalí's *Metamorphosis of Narcissus*, which, incidentally, exerted considerable influence on his friend Jacques Lacan.[46] This is only a sampling of an inventory that could be extended over almost every period and medium of visual art.[47]

In light of this vast visual tradition, it is hardly surprising that literary scholars have found Narcissus to be a convenient point of entry for the introduction of tools of analysis couched in visual terms, like the gaze. A few have even found Lacan here, sometimes unaware that the story is his mirror's ultimate source.[48] Perhaps, however, it should begin to bother us that this supposed parable of vision actually begins with a blind man. Seeking to know the destiny of the son she is about to bear, Narcissus's mother consults Tiresias, who has just been stripped of his sight by Juno but compensated for his loss by Jupiter, who grants him the second sight of prophecy. Narcissus, the blind soothsayer replies, will live to a healthy old age, "if he does not come to know himself."[49] Oracles invite diverse interpretations, and this one has been no exception. Commentators enthusiastically agree that the prediction parodies the inscription over the entrance to the most famous oracle of the ancient world, that of the Temple of Apollo at Delphi: *Gnōthi seauton*, "Know thyself."[50] This allows the episode to unfold as a farce of the Socratic search for self-knowledge.[51] Scholars have differed, however, about what exactly seals Narcissus's doom. Does he "know himself" when he spies the reflection he mistakes for another being, or is the prophecy instead fulfilled a while later, when he begins to understand that what he sees is something he cannot have?

All of this is quite thought-provoking, but in one crucial sense it blunders in the way of most oracular misreadings, seeking to know what the prophet meant before really listening to what he says. For interpreters seem to have taken no note of the fact that Tiresias's words provide the episode's first echo:

> **si se non no**verit

This singsong Tiresias has little precedent in Homer, Sophocles, or Euripides, all of whom invoke him but mostly have him speak fairly ordinary Greek verse.[52] A far closer parallel may be found instead in the *Agamemnon* of Aeschylus, where the ecstatic exclamations of a different prophet, Cassandra, bear a marked resemblance to the repetitive sounds of birdsong[53]—the self-echoing music that, in fact, rings through so many of the myths of the *Metamorphoses*. With his Tiresias, however, Ovid may be thinking not so much of the prophets of Greek epic and tragedy as, rather, of something closer to home (though still in Greek), namely, the alliterative sounds of the prophetic verses of the Sibylline books.[54] These have been suggested as the model for the language of Virgil's fourth *Eclogue*, which begins by calling upon the Cumaean Sibyl and ends with this:

> **Pan etiam, Arcadia** mecum si iudice certet,
> **Pan etiam Arcadia** dicat se iudice victum.
> **incipe, parve puer**, risu cognoscere **matrem**:
> **matri** longa decem tulerunt fastidia **menses**.
> **incipe, parve puer**: cui non risere **parentes**,
> nec **deus** hunc **mensa, dea** nec **dignata** cubili est.[55]

This kind of Sybilline soundplay, coupled with Plato's insistence that all prophecy requires "interpreters,"[56] may be part of what once led scholars to suppose that the Pythia, Apollo's high priestess in his oracular shrine at Delphi, breathed in volcanic fumes and uttered only babble, subsequently converted into words by her attendant priests. That view is flatly contradicted by the evidence, but there was at least a pervasive ancient tradition that the Pythia delivered (or once upon a time had delivered) her prophecies in verse.[57] A stronger indication of a divinatory tension between sound and sense can instead be found at antiquity's second-most famous oracle, that of Dodona, where the whispered will of Zeus was heard in the wind blowing through the god's sacred oak. "The Greek spirit, on the whole, therefore, is free from superstition, since it changes the *sensuous* into the *sensible*," argues Hegel, on the basis of Dodona and the like.[58] Roland Barthes, though, is more compelling, hearing instead a sensuous expansiveness that he memorably calls "the rustle of language" (*le bruissement de la langue*).[59]

In context, however, Tiresias's jingling self-echo must be connected not so much back to his divinatory roots as forward to the sound play of the story whose ending he has just prophesied. This will include, of course, the famously futile echoes directly attributed to their namesake nymph—whose name, by the way, though you might not have guessed it, ultimately derives from the same Indo-European root, **(s)wāgh-*, that gave Latin an onomatopoeic verb, *vagire*, for the repetitive wailing ("wah wah wah!") of infants.[60] Here she begins to throw the youth's final word or words back at him:

> Dixerat "ecquis **adest**?" et "**adest**" responderat Echo.
>
> He said, "Is anybody here?" And "Here!" answered Echo.[61]

And again:

> "Huc **coeamus**" ait, nullique libentius umquam
> responsura sono "**coeamus**" rettulit Echo.
>
> "Come where I am and let's meet," he said, and never to answer a sound more willingly, Echo answered, "Let's meet!"[62]

Her enthusiasm, commentators rush to explain, depends on the painfully comic *double entendre* of *coeamus*: "Let's meet" or "let's have sex." But these sporadic bursts of language are only bubbles on the narrative surface of the episode's even more resonant depths. In the following, the repeated shared stem of *vox* and *vocare* actually drown out the echo itself, which is only implied:

> **Vo**ce "**ve**ni" magna clamat; **voc**at illa **voc**antem.
>
> In a loud voice he cries, "Come!" And she calls the one who calls her.[63]

Echo soon steps out of hiding, and when Narcissus flees her embraces, she rapidly wastes away. *Vox manet*, however: "Her voice remains."[64] In fact, it will haunt even the most visual moments of the story to come. Here desire for his own reflected beauty first seizes Narcissus:

> Se cu**p**it im**pr**udens et qui **pr**obat i**p**se **pr**obatur,
> dum**que petit petitur, pariterque a**ccend**it et a**rdet.
>
> Foolish, he wants himself: the one who likes is the one who is liked;
> while seeking he is sought; in equal measure, he inflames and burns.[65]

His death begins to unfold:

> **Per**que ocul**os per**it **ipse suos**.
>
> Through his own eyes his life drains away.

He recognizes the illusion, though this will not save him:

> "**I**ste ego **sum**! sensi, nec **me me**a fallit **im**ago."
>
> "That's me! I've understood, and my image does not deceive me."[66]

These are only a tiny sample of the story's repetitions (within lines, between them, and across the entire episode), which together offer a kind of phonological symphony that, while consistent with the music of the broader poem, is perhaps unique in its intensity.[67]

In the end, surely what we should be finding here is not so much Lacan as Kristeva, for the episode offers precisely the sort of "deluge of the signifier" by which poetic language overwhelms the symbolic with what she calls the "semiotic," which stands outside language and, simultaneously, is the very stuff of which language is made.[68] (The roots of the Kristevan semiotic are, in fact, in the prelinguistic body and its relationship, *avant la lettre*, to "mama.") Kristeva likens the symbolic to a "sacrifice"; art, however, stands "face-to-face" with that sacrifice and "mimes" it, superabundantly reproducing its signifiers and so the whole "movement of the symbolic economy."[69] "Art—this semioticization of the symbolic—thus represents the flow of jouissance into language," where it becomes audible in, for example, "phonemic devices (such as the accumulation and repetition of phonemes or rhyme) and melodic devices (such as intonation or rhythm)," which constitute what she dubs the "genotext."[70] In the poetic genotext, "poetry—more precisely, poetic language" serves "to introduce through the symbolic that which works on, moves through, and threatens it."[71] Contemplating, with Mallarmé, "the semiotic rhythm within language," she observes, "Indifferent to language, enigmatic and feminine, this space underlying the written is rhythmic, unfettered, irreducible to its intelligible verbal translation; it is musical, anterior to judgment, but restrained by a single guarantee: syntax."[72] In Kristevan terms, Ovid's poem, in its miming of the sacrifice of Narcissus and Echo—both, in slightly different ways, are undone by the symbolic—explodes into a semiotic symphony that makes even the syntactical pillars of Latin vibrate with song. We may partly compare Kristeva's teacher Roland Barthes, who describes a kind of "listening" (in his encyclopedia entry on that rubric) directed not at the "deciphering" of signs but, rather, at "the shimmering [*miroitement*] of signifiers."[73]

Alternatively, we may read Echo as Carolyn Abbate reads operatic heroines, namely, as "woman undone by plot yet triumphant in voice," emblematic, therefore, of music as "sound that seems to signify nothing (and is nonetheless splendid)," destined to be reduced, in our efforts at interpretation, "to words that claim discursive sense but are, by comparison, modest and often unlovely."[74] Starting with feminist readings of opera, but also querying more pointedly musicological perspectives, Abbate seeks to move beyond the critic's seemingly inevitable transformation of music into some kind of "narrative." Opera, she observes, even though partly made of words and (lavish) stories, endlessly resists, from within, its own narration. By way of example, she considers the famous "Bell Song" from Léo Delibes' opera *Lakmé*, which is preceded by "a wash of wordless coloratura" as the title character prepares to sing, both to the scene's internal audience and to the one that listens beyond the orchestra.[75] What follows is "pure voice":

> As if to confirm her own status as sonority rather than story, Lakmé produces music that might itself be regarded as *working against the story she narrates*, since the two musical verses, by remaining similar, by repeating, in some sense deny the progressive sequence of changing events that are recounted in Lakmé's words.[76]

This sounds something like Ovid's Echo, even if Lakmé temporarily succeeds in captivating her beloved. To concentrate on either heroine's failure finally to write the story she wants (losing her beloved, Lakmé ends the opera by poisoning herself) is to miss her triumph, "in which vocal performance will indeed overpower plot," at least for a while.[77]

Have we stopped our ears to such voices, and their shimmering "music," in Ovid? The projected six-volume commentary on the *Metamorphoses* under the general editorship of Alessandro Barchiesi, one of the most influential Ovidian scholars of recent decades, which began appearing in Italian in 2005 and will soon be published in English translation, devotes thirty-two pages (in tiny type) to the 171 lines that comprise the Echo and Narcissus episode—a page, on average, for every four or five verses.[78] Dense as they are, these pages barely mention the ubiquitous sound effects we began to sample above, beyond Echo's repetition of whole words and phrases. The project's single volumes are the products of different hands, but the one in question (volume 2) is the work of Gianpiero Rosati and Barchiesi him j. The omission might at first seem surprising, given that Barchiesi, one of the leaders of what he characterizes as the "striking revaluation of Ovid as a master in the second half of the twentieth century,"[79] expresses regular interest in Ovid's "voices" and even "music."[80] But for Barchiesi and what he has called "my generation of Latinists,"[81] voices are always "speaking volumes," that is, nodes of significance that authorize a project that is extensively and, for the most part, exclusively exegetical.

Sometimes that exegesis follows historicist lines, either in the narrow sense of relating Ovid to Rome and Augustus (the project of Barchiesi's earlier *Il poeta e il principe*, though this is largely about Ovid's *Fasti*), or in a broader sense, such as reading for gender, invoked already in the commentary's general introduction, by the late Charles Segal, and employed repeatedly in the pages devoted to Narcissus and Echo. Most of the time, however, Barchiesi and his "generation" prefer the shadow historicism of "intertextuality," the term they borrowed (from Kristeva, in fact) to rebaptize the ancient literary machine, which they largely treat as an economy of exchanged values (between poets, between languages, among communities of readers). Their goal is to uncover the means of production of ancient literary meaning; words are this economy's coin, and sounds that are less than words can only serve to mystify its mechanisms—unless, of course, they too can be shown to serve meaningful ends, which is what Wills attempts to do in the second half of *Repetition in Latin Poetry*. It is especially this second exegetical mode—intertextuality and the like—which, I would suggest, is beginning to look ever increasingly like the kind of hulking *interpretatio interpretationis causa* against which Susan Sontag cast her brief manifesto, "Against Interpretation," in 1964. Perhaps the time has come again to try to escape the hermeneutic imperative—and in more than just Ovidian scholarship.

The problem of interminable analysis is not that it tells us too much but, rather, that what it does tell us crowds out all perception of what it cannot. Nowhere, perhaps, is this clearer than in the banishment of sound from what scholars have become so fond of referring to as the "world" of ancient literature, or of a particular author, or of a particular work. For it has become commonplace, for example, to assert that Ovid ignores and even scorns Echo as surely as Narcissus does. Echo vanishes, however, only if we perversely insist on *looking* for her; listen, and she is instead everywhere. Curiously, Barchiesi's collaborator Rosati, in a study published three decades before their joint commentary, notes a number of the episode's more pointed sound effects.[82] Nevertheless, he subordinates these, like a kind of underlining, to the optical reflections that are, for him, the real "story" here. All of this is in the service of his overall thesis, namely, that Narcissus represents the very work of poetry: where Alberti saw a painter, Rosati, *ut pictura poesis*, finds Ovid himself.[83]

The more one thinks about it, however, the stranger it becomes that we have been so ready to see Narcissus the portrait-gazer as a poet. One could just as reasonably read him precisely as poetry's antipode. Indeed, the rivalry of the "sister arts" of poetry and painting is nowhere clearer than it is in the epigram that the Late Antique poet Ausonius addresses to a painter in the voice of Echo, who concludes with the taunting advice, "If you want to paint my likeness, paint a sound."[84] The epigram became a favorite in the Renaissance, for which Echo would emerge as a kind of poetic super-Muse.[85] Already in the fourteenth century, Boccaccio had confidently placed her home on Mount Parnassus.[86] As Lynn Enterline deftly observes, already in Ovid's own poem-closing bid for immortality as a disembodied, endlessly repeating voice, he "resembles no character in the poem so closely as his own Echo."[87]

If, in modern scholarship, Narcissus (a kind of painter) must repeatedly be said to figure the poet, why not Medusa (a kind of sculptor)? So asks Victoria Rimell, another voice of dissent against the prevailing picture, noting that Ovid's "poetry encompasses and toys with desires and fears which cannot adequately be contained within the banks of Narcissus' pool."[88] Leaving those banks (and, more broadly, the Narcissus-Orpheus-Pygmalion triumvirate that variously reigns in recent criticism as the poem's man-behind-the curtain) far behind, Rimell offers an array of alternatives, both within the *Metamorphoses* and beyond, simultaneously provoking us to resist choosing this one or the other.

Let us stay poolside a moment longer, though, for there are alternatives here too. Naturally, there is Echo herself, the first of Rimell's "competing models of the artist."[89] (Ovid, by the way, just may have been the first to link Echo and Narcissus, though his influence insured he would not be the last.)[90] Perhaps, however, Ovid's most compelling avatar here is neither of the doomed youths but, rather, the character who sets the story's sounds in motion: Tiresias. It is again Enterline who detects a pointed resemblance between this character and our author: as Tiresias had lived both as a man and as a woman and could speak to the erotic experiences of both, so Ovid, in this and other poems, ventriloquizes male and female lovers with equal ease.[91] At a more basic level, Tiresias shares at least two things with poetry's founding father, Homer: blindness and the title of *vates*, the Latin word that can mean either "poet" or "prophet."

The Homeric poems themselves offer a number of characters that can be read as doubles of the poet or singer; the most striking of these surely are the Sirens, past whom Odysseus's men row, their ears stopped with wax, while their captain, chained to the ship's mast, allows himself to listen:

> ἴδμεν γάρ τοι πάνθ' ὅσ' ἐνὶ Τροίῃ εὐρείῃ
> Ἀργεῖοι Τρῶές τε θεῶν ἰότητι μόγησαν,
> ἴδμεν δ' ὅσσα γένηται ἐπὶ χθονὶ πουλυβοτείρῃ.

> For we know every toil on Troy's wide plain
> by Greeks and Trojans when gods so willed,
> all that happens on the nourishing earth, we know.[92]

What the Sirens know foremost—the story of the Trojan War—is something that Odysseus himself already knows all too well. So do we, if we know the *Iliad*, which in essence is what they propose to sing, again.[93] But repetition (of the story, of *idmen*) turns what we already have heard into refrain, into music. *Idmen . . . idmen . . .* "we know . . . we know . . ." One thing we all ought to know is this: Sirens sing. So, in the ancient understanding of things, do poets: *Arma virumque cano*, "I sing of arms and a man," and so on. Ovid, too, introduces his *Metamorphoses*, in its opening lines, as "song," specifically as a "continuous" one (*carmen perpetuum*). Fifteen books later, just before his final *vivam*, he inserts a parenthetical condition in which he casts himself as a *vates*: he will live "if there is any truth in the predictions" of such (*si quid habent veri vatum praesagia, vivam*). Is this lingering doubt about prophecy or poetry? We may read *vates*, and thus the caveat, either way, hearing a final flutter of meaning as Ovid ends the poem he has built as much of sound as of sense on one last murmuring alliteration: *veri vatum . . . vivam*. Repetitions like these, especially as isolated specimens, have come to make some scholars slightly uncomfortable: they seem like child's play, or indeed, baby talk, which is why they rarely find a place in Barchiesi's grownup commentary. As the building blocks of language, however, they are those of poetry too, assimilated here to the whispered wisdom, often incomprehensible, of the gods.

Actually, this particular alliteration just might earn a nod in Barchiesi's yet unpublished sixth volume (with commentary instead by Philip Hardie), since it can be used to make a point Wills could have but uncharacteristically did not make in *Repetition*.[94] For the alliteration here underscores a clear allusion to a famous phrase of the epitaph Ennius wrote for himself, *volito vivos per ora virum* ("living I light upon the lips of men"), source also of Ovid's *ore legar populi*, scooping up along the way the similarly inspired *victorque virum volitare per ora* ("to light victorious upon the lips of men") of Virgil's third *Georgic*.[95] Ennius, the self-styled "Roman Homer," was, like other early Latin poets, relentlessly (and to hostile ears, excruciatingly) alliterative—so much so that alliteration has been regarded as a kind of inaugural condition of Latin poetry, one that distinguishes it from its far less alliterative Greek models.[96] As a result, to reduce Ovid's (and Virgil's) repeated *v* to allusive emphasis would be to sidestep an older and more basic question of Latin alliteration (and poetic repetition generally), which in this particular Ennian phrase plainly looks not back, but forward, crafting from repetition *in* poetry a figure for the repetition *of* poetry. As is often observed, the former aids the latter by making poems easier to remember, but this longer-term memory and the traditions it makes possible are themselves extensions of more fleeting cognitive processes and sonic pleasures. Every *vates* (Ennius, Virgil, Ovid) must worry first about this smaller picture before trying to situate himself in any larger one: I repeat (*v* . . . *v* . . .), therefore I shall be repeated (*vivam*).

The name of *vates* links Ovid to the legendary first poet, Orpheus, whom he again and again refers to as a *vates*, in the sense of "poet," in the extraordinarily long episode he devotes to him. But the head of Orpheus, after reaching the end of the Hebrus and floating across the sea to Lesbos, is said by others to have turned from poetry to prophecy, becoming a famous oracle.[97] Indeed, *vates* loops the poem's ending above all back to the first person in the poem to be given that title, the prophet Tiresias, and to his own alliterative prediction in the form of a conditional clause: *si se non noverit*. Now that we can hear those words, we can perhaps risk trying to understand them. If knowledge spells doom for Narcissus, and if, in the end, nothing less than the poem's survival is at stake for Ovid, then what exactly are *we* supposed to do to escape our hermeneutic vice? Perhaps just this: step away from the mirror; close our eyes; listen. "In place of hermeneutics we need an erotics of art," proclaimed Sontag, a half-century ago.[98] Let's do it, then, and fall back in love (were we ever really not so?) with Echo.

TRACK THREE

Eine kleine Nachtmusik

Heard melodies are sweet but those unheard
Are sweeter . . .

—JOHN KEATS, "ODE ON A GRECIAN URN"

Propius igitur accessi et "nosti," inquam, "magister, verbum illud scilicet e Graecia vetus, musicam, quae sit abscondita, eam esse nulli rei?"

I came closer and said, "Professor, do you know the old proverb, the one from Greece, that says that hidden music gets you nowhere?"

—AULUS GELLIUS, *ATTIC NIGHTS* 13.31.3

One carefree evening in the middle of the second century CE, a wealthy young Roman citizen from Asia Minor who had been sent to the capital to complete his studies threw himself a birthday party in a country house just outside the city. He was "good-natured—not just rich, but rich in character" (*laetae indolis, moribusque et fortuna bene ornatus*), and he had the good taste to invite to his banquet not only his friends but also his teachers.[1] We learn all this from that amiable antiquarian Aulus Gellius, the author of a freewheeling Latin miscellany of personal anecdotes and scholarly commentary that he called the *Attic Nights*, after the Athenian countryside in which he had first begun to burn the midnight oil over his memoirs. Gellius had come to the party with his own teacher Antonius Julianus, who, as one of the best-known professors of rhetoric in the city, probably counted the birthday boy among his pupils. Gellius may have appeared on the list of friends (depending on the date, the two could have been fellow students), but the slightly forced way he describes being accompanied by Julianus (*venerat tum nobiscum ad eandem cenam*) hints to the reader who knows the ways of his author all too well that it is instead Gellius who is tagging along, uninvited—something he had a lifelong habit of doing.[2] Of course, we ourselves are in no real position to criticize, since we make it through the door and join the festive crowd only because we, in turn, know Gellius.

In his haste to move on to the main event, Gellius barely gives us time to choke down an hors d'oeuvre, relegated to a single subordinate clause: "when the nibbling came to an end and it was now time for drinking and conversation" (*ubi eduliis finis et poculis mox sermonibusque tempus fuit*).[3] Conversation, though, is not what we get, at least not yet, for Julianus insists on some entertainment: "He requested a show of singing and dancing by the expert performers of both sexes that he knew were among the young man's possessions" (*desideravit exhiberi, quos habere eum adolescentem sciebat, scitissimos utriusque sexus qui canerent voce et qui psallerent*).[4] Out they come, to sing "many works of Anacreon, and of Sappho, as well as several sweet and charming love elegies by recent poets" (Ἀνακρεόντεια *pleraque et Sapphica et poetarum quoque recentium* ἐλεγεῖα *quaedam erotica, dulcia et venusta*).[5]

Without warning, we leave the party for a moment and find ourselves with Gellius in his study, where he looks back on the evening's pleasures from a distance that, properly speaking, could be anything from hours to years:

> Among many other things that entertained us were some absolutely delightful verses which were composed by Anacreon in his old age. I have decided to record them here, so that, for their duration, the ceaseless toil of these lucubrations might find a little rest in the sweetness of their words and music.[6]

There is some important play on tense here: Does refreshment come from remembering the verses, from actually writing them down, or from reading them once they have been recorded? Perhaps we can say that this last is mostly meant, for it describes as well our own share of pleasure, given that we are expected to make the *Attic Nights* our own, matching Gellius's late-night writing with our own late-night reading, wherever (and whenever) we are. Indeed, Gellius's digression might be read as a suggestion that the young man's soirée, like all the other chapters of the *Attic Nights*, is secretly a party in our own honor. At the same time, however, the fact that Gellius does not apostrophize us directly (I have recorded this for you, gentle reader) suggests something slightly different, and this, to use a well-worn analogue, is a poignantly personal moment *à la recherche du temps perdu*, in which the anticipation of our greater distance from the events described is only an amplification of the author's own sense of loss:

> And suddenly the memory revealed itself. The taste was that of the little piece of madeleine which on Sunday mornings at Combray (because on those mornings I did not go out before mass), when I went to say good morning to her in her bedroom, my aunt Léonie used to give me, dipping it first in her own cup of tea or tisane.[7]

Gellius-as-Proust gives us no madeleine but only barely remembered *edulia*, "dainties," accompanied not by a teacup but by a wine cup—or to be more precise, a Greek song about a wine cup, quoted in full. The verses are addressed to the god Hephaestus, who is asked to fashion a silver chalice but to decorate it with no weapons or astral scenes but, rather, with grapevines, the fruit of which is to be trampled by figures chased in gold: Dionysus, Eros, and the poet's beloved Bathyllus. In other words, this is doubly a wine cup: both inside and out. Indeed, we may even say that it is trebly so, for Anacreon himself was famously bibulous and even was commemorated, it is recorded, by a statue showing him old, barefoot, and two sheets to the wind, singing about Bathyllus or some other beautiful boy.[8] Of course, a skeptic could object that a poem about a wine cup depicting winemaking, even when written by a drunken poet, will not so much as wet your whistle. In fact, despite the promise of imminent drinking, our own cup will remain empty, for the only things really on offer at this banquet, at least by the time we get there, are words.

The thirsty reader's frustration is the fruit of a rather remarkable game that Gellius is playing, consciously or not, with the limits of memory and language. Indeed, the chapter he devotes to the party offers a steady escalation in what it records. This begins with the food and drink, which Gellius notes only in passing. Next enters the troupe of performers, *scitissimi utriusque sexus*, "expert" and "of both sexes."[9] The latter detail presumably is meant above all to help us imagine their voices, but attention to their gender joins their dancing and their partly erotic subject matter to conjure a spectacle that is more than innocently artistic.[10] Perhaps Gellius even plays coyly on the colloquial use of *scitus* to mean "good-looking": these are "very fine" musicians in more ways than one. Gellius next gives us some generic information—authors, themes—about what they sang; then comes the verbatim transcript of the wine-cup song (not really by Anacreon, but an imitation from the corpus now known as the *Carmina Anacreontea*):

> Τὸν ἄργυρον τορεύσας
> Ἥφαιστέ μοι ποίησον
> πανοπλίας μὲν οὐχί·
> τί γὰρ μάχαισι κἀμοί;
> ποτήριον δὲ κοῖλον
> ὅσον δύνῃ βάθυνον.
> καὶ μὴ ποίει κατ᾽ αὐτό
> μήτ᾽ ἄστρα μήτ᾽ Ἀμάξας·
> τί Πλειάδων μέλει μοι,
> τί δ᾽ ἀστέρος Βοώτεω;
> ποίησον ἀμπέλους μοι,
> καὶ βότρυας κατ᾽ αὐτῶν
> καὶ χρυσέους πατοῦντας
> ὁμοῦ καλῷι Λυαίῳι
> Ἔρωτα καὶ Βάθυλλον.

Shape now your silver,
Hephaestus, and make me
not armor (with battles
what have I to do?),
but hollow a wine cup,
as deep as you're able,
on which please do not make
the stars or Big Bear.
What have I to do
with Pleiads or Ploughman?
But make me some grapevines
with bunches of fruit.
Then add some who trample
the grapes with fair Bacchus
and let them be gilded:
Bathyllus and Love.[11]

At this point, however, we are only halfway through the chapter, for our compensatory draught of words is immediately followed by the conversation that had been delayed by the concert and which will occupy the remainder of the evening.

Here, at last, Gellius is truly in his element, for reporting the wise and witty discourses of his friends and acquaintances is the most regular business of his *Attic Nights.* Tonight's table talk is set in motion by a gang of rowdy Greeks looking for some good-natured trouble. They decide to pick on Julianus, calling him a Spanish bumpkin, less an orator than a "screecher" (*clamator*), and a raving professor of a language to which "the charm and sweetness of Venus and the Muse" are utterly alien (*eiusque linguae exercitationes doceret quae nullas voluptates nullamque mulcedinem Veneris atque Musae haberet*).[12] What, they jeer, could he possibly have to say about Anacreon and his peers? Surely no Latin poets provide comparable delight. Maybe a few things by Catullus, or Calvus—but nothing else.

Julianus takes the bait and, indignant (or pretending to be), prepares to defend his mother tongue as if fighting for his country's "altars and hearths."[13] It is true, he admits, that Greece reigns supreme in artistic decadence and frivolity. "But before you convict us—that is to say, the Latin name—of insensitivity to love, like a bunch of uncivilized brutes, let me cover my head with my cloak, something they say that Socrates once did in order to deliver some slightly racy material, while you listen and learn..."[14] Julianus then performs pieces by four Latin poets, each also transcribed by Gellius, after a brief aside in which our author gushes that "in Greek or in Latin, I think that nothing more elegant, charming, polished, or pure can be found."[15] The chapter abruptly ends after the fourth erotic poem, without any word on whether the Greeks were convinced. Given that none of them seems to have noticed that the final piece, presented by Julianus as being by Quintus Catulus, was actually a translation of a Greek original (something Gellius never realized, though the same cannot be said of later, gleeful commentators), it may well be that, by this point in the bibulous party, no one was paying very much attention.

This entire evening would barely be worth remembering were it not for the fact that, in the retelling, Gellius three times uses a word crucial to the broader investigation we have undertaken in this book: *vox*, "voice." The first and third instances depend on another important word, *canere*, which Latin students no sooner learn to translate "to sing" than they are told to be careful: *Arma virumque cano*, "I *write a poem* about arms and a man..."[16] The semantic range of *canere*, however, extends well beyond the conceit of the poet as bard. On the one hand, it can refer to the musical sounds produced by any instrument—pipes, horns, strings. This usage prompts Gellius to describe what the young performers do by the slightly odd expression *canerent voce*: they "make music with the voice." He means only that they sang and soon will switch to the ordinary use of *canere* alone, but the point of his initial phrase is to contrast the rare pleasure of these fine instruments with the dime-a-dozen distractions of dancing flute-players. On the other hand, *canere* can designate ancient performances that, could we hear them today, we might hesitate to place in the category of song proper. Such is the case when Gellius tells us that Julianus "sang" (*cecinit*) a series of poems "in a voice as sweet as it gets" (*voce admodum quam suavi*).[17] The verses in question are in elegiac couplets, not generally thought to be musical in the way of lyric poetry; Julianus's delivery belongs instead to the art of *recitatio*, the expert "reading" of literary texts.

To all this "singing" the party inevitably added a generous dose of the kind of vocal performance most on display in the *Attic Nights*: oratory, the art Julianus taught and for which the rambunctious Greeks derided him as a "screecher," ultimately eliciting from him a rejoinder that, as we have seen, eschews extended declamation for the surprise twist of a poetic recitation. To state the obvious, what all of these modes (oratory, singing, recitation) have in common is their reliance on the voice. Our efforts to join the party with Gellius might well lead us therefore to describe the voice as that quality of a vocal performance that can be described ("skilled and of both sexes"; "a screecher"; "in a voice as sweet as it gets") but not transcribed, the latter possibility being reserved for words themselves, not for the ways in which those words were delivered. Much like the wine we are never allowed to taste, voices, we might sigh, are another treat that Gellius cannot save for us.

Consider, however, the account's remaining instance of the word *vox*, which comes in a passage we already have considered, as Gellius introduces the sung verses he supposes to be by Anacreon:

> Oblectati autem sumus praeter multa alia versiculis lepidissimis Anacreontis senis, quos equidem scripsi, ut interea labor hic vigiliarum et inquies suavitate paulisper vocum atque modulorum adquiesceret.
>
> Among many other things that entertained us were some absolutely delightful verses which were composed by Anacreon in his old age. I have decided to record them here, so that, for their duration, the ceaseless toil of these lucubrations might find a little rest in the sweetness of their words (*voces*) and music.[18]

Here, as often, *vox* (plural, *voces*) means not "voice" but "word." That the Latin word for "voice" would also be used as a word for "word" is not in itself surprising—not even in the case of written words, given the alphabetic West's long understanding of these as images of speech. This extended sense, however, has led to peculiar juxtapositions like those of this particular chapter in Gellius, where *vox* has two meanings that are not only different but, seemingly, mutually exclusive, designating either exactly that which he can write down, or exactly that which he cannot. Indeed, in English translation (though not in other languages closer to Latin), we are obliged to impose two words—respectively, "word" and "voice"—where the Latin text has only one.[19]

If the categories seem to us distinct, our own memories of sung words and singing voices sometimes blur. Consider the following lyrics:

> Don Giovanni, a cenar teco
> m'invitasti e son venuto!
>
> Don Giovanni, you invited me
> to dine with you, and I've come!

No one who does not recognize these words could ever guess that they corresponded to the extraordinary music notated, in its composer's own hand, in figures 4 and 5. By contrast, devoted opera lovers will at once identify the words of this invitation as those thundered by the Commendatore, who, risen from the dead, has just appeared at Don Giovanni's front door, ready to send the serial seducer to Hell and to bring Mozart's opera to an end (almost). Moreover, when they read those words, they not only can imagine they hear the music notated in the score, but they may even supply additional musical details not contained therein: the sound of a particular orchestra, the tempo of a particular conductor, the voice of a particular bass, either recalled from an especially memorable performance (or synthesized from a lifetime of such) or known from a favorite recording, perhaps played hundreds of times.

In other words, not even the most sophisticated system of notation can "record" for an expert (but inexperienced) reader all that a

Figures 4 and 5. Wolfgang Amadeus Mozart, *Il dissoluto punito, ossia, Il Don Giovanni*, Autograph Manuscript, Bibliothèque Nationale de France, Département de la Musique, MS. 1548 (7), ff. 22r-v. Reproduction courtesy of the Bibliothèque Nationale de France.

e son ve-nuto
non l'avrei giammai creduto,
ma farò quel che po-trò! Leporello un'altra

few words can prompt even an inexpert (but experienced) listener to remember. Just what a musical text represents depends on what you already know. For readers who know music in one sense—those who can read music but who have not ever heard Mozart's opera—the score conjures the right notes, phrasing, and instruments. For readers who know music in another sense—those who cannot read music but who do know the opera—the lyrics serve the same basic purpose (with the bonus of additional musical detail). For readers who know music both ways, even the score accesses not only notational literacy but also musical memories potentially shared with those who instead cannot read the same.

The phenomenon is hardly limited to opera. Most readers would be unable to read the lyrics of "Stairway to Heaven" without hearing some very specific music, and in all but a very few cases, this will not be remembered from Dolly Parton's 2002 cover of Led Zeppelin's greatest hit.[20] In the instance of these likewise famous lyrics, one medium (the record album) has trained us to read another (the written text) in a particular way. "Stairway to Heaven" also offers a good example of how, in the phonograph's wake, it has become increasingly possible for a piece of vocal music to be associated, from the get-go and almost indelibly, with the identifiable sound of a specific human voice, to the extent that, for some listeners, Parton's version is not really the same "music" at all, not only because she made changes to the music itself, but also, and above all, because her voice is decidedly not Robert Plant's. Indeed, in the mind of the diehard Led Zeppelin fan, Parton's recording, though it belongs, in technological terms, to a more advanced medium than that of a text, may seem to offer less "fidelity," as a record of the song they know, than do the original album's liner notes or any other text of the lyrics. Crucially, this prejudice for the text is born not so much of a sense that the words are more important than how they are sung as it is of the opposite view. Rival recordings, by contrast, potentially corrupt not so much the words as the sound.

My examples have sought to show how, even in the presence of ostensibly superior technologies, words can at least seem to record the full range of musical experience, up to and including the voice. In each case, of course, they do so by accessing musical records in another medium, namely the listener's memory; we have *learned* to associate this text with that music. This association, in other words, is a matter of convention, and it would seem to offer all the arbitrariness of the linguistic sign: beyond the rhymes and refrains that may point vaguely to a corresponding musical shape, there is no way to deduce (musical) signified from (verbal) signifier. They go together simply because we have heard them together, usually many times. Indeed, our two musical examples from Mozart and Led Zeppelin embrace this arbitrariness, going about as far as song can go in turning its back on words. Beginning with its overture, as with the opening notes of our excerpted scene, *Don Giovanni*, like all operas, offers unvoiced as well as voiced music; throughout, the voice is often deployed as an instrument far more of sound than of sense; conversely, the music, sung or not, sometimes does all the storytelling, as is the case in the Commendatore scene, which is hardly about dinner; add to this that Mozart, most of whose works were instrumental, is often said to treat his (other) instruments as (wordless) voices. "Stairway to Heaven" begins with an acoustic guitar, before turning to the singing we have been remembering, which itself eventually yields to a famously elaborate electric guitar solo; the entire composition would become the best-selling single piece of rock-and-roll sheet music ever published; guitarist Jimmy Page would eventually perform an instrumental version of the whole thing when touring alone. Both pieces, in other words, seem to stage their own hybridity: music emerges from its wordy cocoon—and then takes flight.

Our attachment to this metamorphic tale, starting in the nineteenth century, led us to read its larval stages back into earlier epochs and to craft thereof a master narrative of Western musical history as a long struggle to free song from speech and music from song. In this view, the ultimate goal is taken to be not only instrumental but "absolute," that is, disencumbered of any and all semantic obligations.[21] Ancient Greece's embryonic system of musical notation (which may, in fact, have been rather more than embryonic) is taken to be an early step forward. Saint Augustine's anxieties about music, predicated on his observation of how contemporary musical practice was interfering with the understanding of sung words of scripture, give sure signs that progress was being made. Medieval neumes and their successors, against a backdrop of sometimes incomprehensible polyphony, provide the decisive chapter from which there can be no turning back: Beethoven, we are assured, was now inevitable. The account I am caricaturing is not without truth or value, especially when used to map tensions over briefer periods; it is especially illuminating of the medieval notational developments on the basis of which, hardly by accident, it largely has been crafted. As a big picture, however, it has almost too many problems to list. Among other things, if music has always been yearning to escape words, can it really have taken the "harpists" of Early Cycladic art and their successors on the strings four thousand years to compose a sonata or to take their instruments on solo tour (fig. 6)? Freedom was there all along, just beneath their large noses. (In fact, instrumental music has always had its own traditions independent of song, including in antiquity;[22] the teleological view winds up shortchanging even its hero.) To be clear, no serious musicologist today espouses such an epic vision of musical history, at least not without considerable nuance. What makes this story interesting, however, is its implicit and sometimes explicit grounding in the history of musical media. For something similar is going on in our retrospective understanding of the components of song.

Regarding the representation of song, we are by now separated from antiquity by not one but two watershed technological innovations: the medieval development and widespread adoption of modern musical notation, and the nineteenth-century invention of the phonograph. The former technology, even more than our instrumentalist, is the not-so-secret hero of the teleological tale we have just rehearsed. Combined with the latter, it then yields our seemingly self-evident tripartite division of the components of song along the lines of vocal media, that is, into lyrics (what can be recorded by written texts), music proper (what requires instead

Figure 6. Male Harp Player of the Early Spedos type, Marble, Cycladic, 2700–2300 BCE, Getty Villa, Malibu, 85.AA.103. Image courtesy of the J. Paul Getty Museum.

musical notation), and the voice, understood precisely as those qualities of a vocal performance which elude the first two and which can instead be captured only by the phonograph and its successors. (Let us set aside, for the purposes of the present exercise, any instrumental "accompaniment.") The apparent result, from our vantage point, is a simple sum of three independent values: WORD + MUSIC + VOICE = SONG. Recent technologies capture all three and thus provide (or at least approach, within the limits of their "fidelity") a perfect mediation of song: MP3 = WORD + MUSIC + VOICE = SONG. Earlier representations, however, suffer by comparison. A score is SONG – VOICE. Lyrics, worse still, are SONG – VOICE – MUSIC.

What this calculus fails to account for, however, is our actual usage, over time, of the word "song" in relation to media. MP3s give us songs, as once did 45 RPM records. Still, we have not entirely lost the habit of saying the same about sheet music, suggesting that, sometimes, our math must admit a reduced formula: SONG = WORDS + MUSIC. Our problems mount when we reach back to uses that have instead become mostly alien to us. To take the example most germane to our present purposes, the Latin word for "song," *carmen*, regularly refers to a written text, seemingly reducing our formula to, simply, SONG = WORD. How to make this add up? The normal solution has been to doubt the simple equivalence of *carmen* = SONG: when applied to literary texts, *carmen*, like its verbal cognate *cano*, as we already have seen, is made to refer not to song but to "poetry." A mild form of this correction sees the appeal to "song" here as either metaphor (poetry is like song) or metonymy (song is instead an exemplary kind of poetry). A stronger form observes that *carmen* can refer to such a range of utterances (chants, incantations, hymns, poems) that is best understood as a generic designation of "marked speech."[23] There is merit to each of these corrections, but either the latter must lead us to conclude that Latin had no specific word for song (since *carmen* has no clear competitor in this regard[24]), or the former must persuade us that, though there was such a word (i.e., *carmen*), Latin writers, when they say "song," usually mean something else.

A better solution would follow the lead of Gregory Nagy, who, regarding the analogous problem in the ordinary Greek word for "song" (*aiodē*), which Latin usage partly parrots, recognizes that, over time, the Greek word worked both to distinguish song-as-marked-speech from ordinary speech and to distinguish song-as-singing from unsung poetry.[25] This last category, poetry, thus emerges both *as* and *in opposition to* "song," an emergence that Nagy pointedly segregates from the history of writing. Indeed, the watershed, in his account, is instead the adoption of unmelodic (or, at least, less melodic) and unaccompanied "recitation" of certain kinds of verse, including the dactylic hexameter of the Homeric epics, the actual singing of which began to recede into an ever more distant past. Nevertheless, even if, as Nagy suggests, this shift from singing to rhapsodic recitation was, in origin, independent of the rise of writing, over time the two transformations must have met each other half way, decoupling Homeric song from the lyre and rerooting it fully in what Nagy himself recognizes as the ultimate basis of all ancient singing: language.

What happens if we take seriously the proposition that a sequence of (written) words can *be* a song, not just in the broad sense in which Homer's poems continued to be "songs" for the Greeks even after they stopped singing them, but even in cases like that of the unquestionably sung song quoted by Gellius? (For now, let us say that "being" a song means more than merely reminding us of one, as might be said to be the case for our lyrics from Mozart and Led Zeppelin.) There would seem to be two ways to make sense of such a proposition. One would be to suppose that the additional musical and vocal values, like the kinds we have learned to capture with other media, while present in the ancient performance of song and capable of being appreciated in their own right, were so much less important than words to ancient listeners that their absence, like scratches on a vinyl album, compromised the "fidelity" of a verbal record too little to undermine its basic integrity as a representation of song. There may be a grain of truth in this view, especially in comparison to our own musical culture, but it is hard finally to reconcile with ancient celebrations of performance like what we find in Gellius. A somewhat different solution would be to suppose that ancient listeners did instead regard those musical and/or vocal values as indispensible to song, but that they somehow believed that the verbal record itself embodied these to a sufficient degree to qualify as a representation of the complete article. This is the hypothesis I wish now to pursue, using the verses quoted by Gellius for illustration.

Let us start with the best-known "musical" quality of ancient poetry, namely, the quantitative meter to which its constitutive syllables must correspond. This meter is built of patterns not of stress accent but of long and short time values; normatively, the long value was twice as long as the short value, which allows the convenience of analogizing them to modern musical durations.[26] For example, a relatively rare pattern, ionic dimeter, named for the region of Anacreon's birth, combines two short syllables with two long ones and then repeats the sequence:

♩ ♩ 𝅗𝅥 𝅗𝅥 ♩ ♩ 𝅗𝅥 𝅗𝅥

(The final syllable is properly a special case, but for simplicity's sake we may treat it as long.) Switching the two middle "notes" (a phenomenon called anaclasis) yields instead a line that came to be known as an "anacreontic":

♩ ♩ 𝅗𝅥 ♩ 𝅗𝅥 ♩ 𝅗𝅥 𝅗𝅥

Finally, chopping off the first of these (acephaly) and allowing the first value to be either long or short produces what is generally called a "hemiamb":

♩ 𝅗𝅥 ♩ 𝅗𝅥 ♩ 𝅗𝅥 𝅗𝅥

or

𝅗𝅥 𝅗𝅥 ♩ 𝅗𝅥 ♩ 𝅗𝅥 𝅗𝅥

The historical Anacreon did not limit himself to these meters, but his imitators in the *Anacreontea* stick almost entirely to anacreontics and hemiambs,[27] clearly regarded as the poet's metrical signature. The wine song is composed entirely of the latter, at least in the version given by Gellius. Here is its first line:

τὸν ἄρ- γυ- ρον το- ρεύ- σας

There can be little doubt that the song heard by Gellius conformed to the quantities already indicated by the poem's syllables. "Every Greek poet was his own composer, and no poet would write words in elaborate metrical schemes merely to annihilate and overlay these by a different musical rhythm," notes A. M. Dale, quoted with approval by Martin West in *Ancient Greek Music*, adding, "In the surviving fragments of poetic texts furnished with musical notation, the note values are commonly left unspecified, and this is because they were felt to be sufficiently indicated by the metre of the words."[28]

This brings us to pitch, less definitively encoded in the single syllables of our text by two different systems. The first is often overlooked and regards the role of pitch in the pronunciation of vowels (and voiced consonants), an extraordinarily complex matter that can be boiled down, for Greek as for English, to this: *ceteris paribus*, every vowel has a distinct pattern of pitch. Say the following normally, and you should hear a descending "scale": beat, bit, bet, bat, boat, bought.[29] The relationship of these patterns to those of song has mostly been considered by scholars of folk music; a curious recent study shows a close match in the case of "nonsense" singing like yodeling, suggesting that we might be able to look for some of the same wherever the pressures of sense are relaxed to those of sound, something we might well regard to be a foundational condition of poetry and song.[30] To these patterns, Ancient Greek adds a system of rising and falling pitches they called *prosōdia*, a "singing accompaniment," later marked by diacritical marks (rising or "acute," falling or "grave," and "circumflex," indicating a combination of the two).

This system of pitch "accent" has long led to the slightly sensational characterization of Ancient Greek as a "musical language," whether sung or spoken[31]; spoken Latin accent, like our own (and even that of later Greek), is instead marked principally by stress. Indeed, it is possible to speak of a whole system of "melodic accentuation" in spoken Ancient Greek, working not just within words but also across phrases, which in turn provides what has been called the "irreducible element of melody" in Greek poetry and song.[32]

If we add these two pitch-based systems to the metrical one, how close is our text to embodying the sum total of the concert's music *stricto sensu*?[33] Even supposing that a narrow answer were theoretically possible—that is, one that did not require reconstructing the perceptions, both general and idiosyncratic, of the concertgoers, whose musical expectations were entirely different from ours—it is unlikely that we will ever know it; this is in spite of recent gains, through study of surviving notation, in our reconstruction of what at least some Greek music may have sounded like.[34] The best that we can do is to remind ourselves of a range of possibilities. The existence of Greek notation confirms, where it is present, a significant role for extralinguistic pitch, as do, more generally, ancient discussions of music, including those of the surviving technical treatises.[35] But in a text for which we have no notation (and no reason to assume such ever existed), we cannot know to what extent singing pitch beyond linguistic pitch (and elaborations thereof) was still a product of conventions implied by the text itself. Anacreontic songs were some of the most popular pieces in the ancient repertoire. Might they (or "Ionic" songs generally) have been sung in relatively predictable ways? Might even more specific conventions have been attached to Anacreontic hemiambs? Finally, might the Anacreontic wine-cup song, two other versions of which survive from antiquity, be even more tightly constrained by a particular singing tradition? The fact that those variants include a range of metrical changes keeps us short of the situation we found in the familiar lyrics of Led Zeppelin; that is, Gellius seems not to be saying, "Here are the words to a song you all surely know."[36] Nevertheless, the basic mechanism—text signifies music—would be the same. To all of this add other constraints, like those of the ancient *harmoniai* or "modes," which limited the pitches available for melody. Anacreon himself, we are told, restricted his songs to the three most venerable of these (the Dorian, Lydian, and Phrygian); presumably these persisted (perhaps in reduced number, as was the case with Anacreon's meters) in the Anacreontic tradition; this would have provided basic pitch parameters and, perhaps, something more, since, as a leading scholar of Greek music suggests, "there may have been particular melodic formulae associated with particular modes."[37] To be clear, such conventions may well have changed between Anacreon's day, the composition of our song, and its Roman performance, but Gellius may not have been aware of this and in any case could expect common ground with his contemporary readers. So our bottom line is this: between the music "in" the words and the music that might be expected from a particular kind (in various senses) of text, we simply cannot know, for the ancient reader, how much music is actually missing here. The specific musical term used by Gellius to introduce his transcription, *moduli*, which can refer to units of either pitch or duration, might seem to take a narrow view of things, but it is probably only a preciously diminutive metonym (compare the "brief" time it takes to read the quoted *versiculi*) for a delicate little song of which Gellius invites his readers to appreciate every tiny, incremental detail.[38]

Before turning from music to the audibility of the voice in this transcription, let us revisit the measure of the latter in our modern comparanda. The phonographic record and its successors have been key to our age's version of the cult of the *individual* voice, to the point that the commodified voice of a singer has become paradigmatic of our general view of voices as aural signatures. Only a technology capable of distinguishing between Robert Plant and Dolly Parton, for example, now impresses us as having captured their voices. Plant and Parton's voices, however, differ in terms not only of sonic irreducibles but also of broad categories of gender and genre, nature and nurture—male, English, rock, versus female, American, country—that are at least as important to our surprise, when one voice is substituted for another, as is any more specific identity. Such general distinctions can not only be noted but even, in some cases, notated. Thus, the score for *Don Giovanni* gives us eight voices: three sopranos, a tenor, and four basses. There are basses and there are basses; indeed, one bass must play Mozart's title character and another, the Commendatore (though at the premier and often subsequently, the latter bass also sang the part of Massetto). Which difference is greater, however: that between Nicolai Ghiarov and Franz Crass, Don Giovanni and the Commendatore, respectively, on my LP of the opera, or that between either of these and tenor Nicolai Gedda (Don Ottavio), or even soprano Mirella Freni (Zerlina)?[39] Indeed, the operatic conventions exemplified by Mozart reveal in an especially pointed way what is so strange about taking the singing voice as paradigmatic of voices generally, since professional singers like these are the products of training, the first goal of which must necessarily be that of *minimizing* variation: in more ways than one, each of these singers has learned to play the part.[40] While it surely is true that some people were born to sing on the stage, others, in the shower, what distinguishes one professional player from another of the same part must too be attributed to art as much as to instrument—with even the latter partially shaped by training and physical discipline.

Returning to Gellius, we find some points of contact regarding the role(s) of the voice. First and foremost, he tells us that the ensemble included both male and female voices. This does not guarantee, of course, that they all sang each piece. Indeed, solo singing would be the mode most natural to the primarily monodic Anacreontic tradition, with its notional origins in symposiastic or komastic improvisation and its pointed reliance on the first-person singular, as here.[41] Probably, therefore, Gellius expects it to be obvious to us that this piece was sung not by all, but by one particular male soloist who plays the first-person part of the narrator. More telling is the superlative applied to the entire group. By relating that the performers were "expert" (*scitissimi*), Gellius might at first seem to be augmenting our regret at not having been there, but he really is offering us a cue: if you yourself know the rules, then you know how they sang. He is doing the same thing when he tells us that the recitational voice of Julianus was "as sweet (*suavis*) as it gets."[42] Such "sweetness" or "pleasantness" is less some innate vocal quality than it is its opposite: the result of a lifetime of training obtrusive edges away from his own voice and those of his students (albeit with an emphasis on oratory—Julianus is surprising us all with additional skills here). Never been lucky enough to hear a Julianus? Well then, imagine the sweetest reciter of Latin poetry you ever have heard, and bump that voice up a notch or two.[43] Needless to say, Gellius is offering his imagined readers, starting with the sons he mentions in the collection's preface, an additional challenge, for even though silent reading was hardly uncommon in antiquity,[44] there can be little question that we implicitly are invited to play a vocal role in the case of the quoted passages in this chapter. Readers who are *scitissimi* in their own right are the appropriate media players here: the fidelity of what Gellius has recorded really depends on *you*.

There is a further voice here, one that likewise speaks both to and through the taste and erudition of the reader. This is the voice of the poet. "Voice" in this usage is deep-seated style: though it may grow and change over time and across genres, it ultimately suggests to us a unity. Not every voice, however, has only one poet, a curiosity of which the Anacreontic tradition provides dramatic illustration. The *Anacreontea* are the products of many hands spanning perhaps as much as a millennium after the death of the sixth-century BCE poet who lent them not just a name, but signature style and substance. Indeed, the corpus belongs to one of antiquity's most persistent efforts to distill and rebottle a poetic essence, not just as more poetry, but also as spurious biography and iconography based largely on the poems themselves. One could "know" a great deal about Anacreon in antiquity: how much he drank; the fact that he said that boys were his gods; that he lived to age eighty-five, still at it, both as poet and as pederast; what he looked like (he appears on numerous Greek vases, often captioned with his name) and the rock-star clothes he wore (which the Athenians would have read as flamboyantly Eastern); one famous statue, of which a spectacular Roman copy survives, even represented him infibulated, that is to say, with his penis temporarily tied up, possibly in an attempt (futile?) at sexual abstinence, perhaps for the sake of his voice.[45] Of course, one could also know what, drunk and in love, he sang. To write under his name was to assume this largely invented persona—and to make its voice your own. But even though largely ventriloquized, "Anacreon" offers a remarkable parallel to the big-name superstar singers of our own day, whose real and fictitious lives and bodies are likewise known in detail. We think of this phenomenon as distinctly modern, but as part of the signature "branding" of song, it is as old as antiquity.

We have by now turned up the volume quite a bit on Gellius's turntable—though our own ability to play his Anacreontic single will forever remain limited. But I want now to try to listen for something that, though quieter, may prove easier for us to hear, even at this great distance. Once again, however, we shall begin in the more recent past, this time with the help of Roland Barthes, who, in his 1972 essay "The Grain of the Voice," explains his dislike for the then ubiquitous German baritone Dietrich Fischer-Dieskau.[46] The "grain" of his title, which Barthes finds instead in Charles Panzéra, is "the body in the voice as it sings," and this is not some innate timbre or other preexisting condition but belongs, for Barthes, to the consummate artistry of that singer who has learned to treat music not as a product, but as a process, "where melody explores how the language works and identifies with that work." He continues, "With FD, I seem only to hear the lungs, never the tongue, the glottis, the teeth, the mucous membranes, the nose. All of Panzéra's art, on the contrary, was in the letters..." Barthes means French letters; his article takes French aim not only at a German singer (though our hero Panzéra was born Swiss) but at German *Lied* and the internationalization of Western music generally, to which he opposes the rapidly fading strains of French *mélodie*, which "has little to do with the history of music and much with the theory of the text." Here "the *grain*, the grain of the voice when the latter is in a dual posture, a dual production—of language and of music" marks "the very precise space (genre) of *the encounter between a language and a voice.*"

Borrowing from Julia Kristeva's complex binary of "phenotext" and "genotext,"[47] Barthes makes an analogous distinction between song-as-being and song-as-becoming: Fischer-Dieskau offers a perfect realization of the rules and expectations of the "phenosong"—"all the phenomena, all the features which belong to the structure of the language being sung, the rules of the genre, the coded form of the melisma, the composer's idiolect, the style of the interpretation: in short, everything in the performance which is in the service of communication, representation, expression, everything which it is customary to talk about," and so on—but only in Panzéra do we hear the messier polyvalence of "genosong," "the space where significations germinate 'from within the language and in its very materiality.'"[48]

In this attempt to define his elusive quarry, Barthes forges a doubtless unconscious link to ancient seekers of the same, raising the question, I would suggest, of whether "the grain of the voice" might not be an elegantly pleonastic and almost musical name for the voice itself. For Barthes's division, like Kristeva's, bears a remarkable resemblance to a persistent ancient double definition of the voice that by Late Antiquity had become an almost obligatory point of departure for the study of language. That definition's ultimate Latin source seems to have been the first-century BCE grammarian Varro, who himself borrowed the concept from the Stoics. The most distilled version of the definition opens the *Ars maior* of Donatus, written in the fourth century CE and destined to provide the Middle Ages with one of its principal grammatical textbooks:

> Vox est aer ictus, sensibilis auditu, quantum in ipso est. Omnis vox aut articulata est aut confusa. Articulata est quae litteris comprehendi potest; confusa, quae scribi non potest.
>
> A *vox* is air that has been struck and which, barring other factors, can be heard. Every *vox* is either *articulata* or *confusa*. A *vox articulata* is one that can be captured by the letters of the alphabet; a *vox confusa* is one

that cannot be written down.[49]

At first glance, it is quite tempting to translate *vox* here as "sound," especially since another grammarian reproduces the same definition but adds *sonus* (the ordinary Latin word for "sound") as an alternate term: *Vox sive sonus est aer ictus*, etc.[50] *Vox articulata* would thus be "articulate sound"; *vox confusa*, "confused sound," or better, "pure sound."[51] The fact, however, that the same grammarian continues to add *sive sonus* each time he uses *vox* suggests he knows that his gloss is somewhat tendentious. There is precedence for such an equivalence; indeed, yet another grammarian who takes up the same definition actually complains that classical authors are ready to treat any *sonus* as a *vox*.[52] Nevertheless, the use was actually rare and mostly poetic; his own example is from Virgil, describing the waves of a groaning ocean as *voces* shattered against the rocks of the Sicilian shore. Back on the dry land of everyday Latin, *vox* is usually restricted instead to sounds made, in one way or another, by humans or animals. *Vox* does turn up now and then as a translation for Greek *phōnē* when the latter seems to mean little more than *sonus* (which it does, in preference to Greek *psophos*, somewhat more often than its Latin counterpart), and this, of course, is largely what is happening here, with Latin grammarians importing Greek typologies. Rather than conclude that all these words simply mean "sound," we should ask why both languages, despite other options, continued to theorize sound with words that usually referred to the voice.

Indeed, the nature of the beast, as it were, is made abundantly clear by the examples that the grammarians give to illustrate the category of the *vox confusa*. Far from embodying "pure sound," most of these examples seem to have been chosen precisely for the way in which they approach without actually reaching the linguistic status that would assign them instead to the category of the *vox articulata*. Two common examples are found already in Varro: the whinnying (*hinnitus*) of a horse and the bellowing (*mugitus*) of a bull.[53] Not only are these animals potentially trying to tell us (or each other) something, but the sounds they make, which we have just been told cannot be written down, are here in some general sense in these very words, *hinnitus* and *mugitus*, both of which demonstrate what we might call onomatopoeia or, in the technical terminology of linguists, "iconicity"—that is, they are signs that embody actual properties of the thing they signify.[54] Later writers mine the same vein, adding the *rugitus* (roaring) of wild beasts and the *sibilus* (hissing) of snakes, or direct transliterations of animal sounds, like *coax*, the croak of a frog, famously used by the chorus of frogs in the comedy of Aristophanes named for them.[55] Another example, the *rabies* (rage) of dogs skips the onomatopoeia and goes directly to what is being expressed. A final example is birdsong (*avium cantus*), the zoological sound that has most often been assimilated to human speech and even music, from Aristophanes to Ovid and beyond.[56]

Human music provides other examples: the sound made by a simple pipe (*fistulae auditus*) and, returning to onomatopoeia, the *tinnitus* (clanging) of cymbals. But these are relatively crude instruments; another grammarian chooses instead the *tibia* and the organ but thus finds himself compelled to class their sound as *vox articulata*, for which he then creates a special subcategory of musical sound, which can be "understood" (*intellegitur*) even if it cannot be transcribed. Outside music, a range of other human sounds count as *vox confusa*: whistling or hissing (*sibilatus*, seen before for snakes), groaning (*mugitus*, seen before for bulls), laughter, cheering, and finger-snapping; to these we can probably add the *stridor* listed by one grammarian, which can refer to any high-pitched sound but here likely designates a human or animal "screech." A tool beyond the body itself is needed to produce a final human sound, the cracking (*strepitus*) of a whip, which completes our picture of the communicative *vox confusa* with a message all too easily understood by man and beast alike.

Of the grammarians' seventeen examples, only the two that remain lack immediate human or animal agents. One, *casus ruinae*, is a somewhat odd phrase for the crash of something collapsing, which sometimes, at least, will be the work of human mischief. The last, *undarum pulsus*, "the pounding of waves," sounds suspiciously like an echo of Virgil's personification of the sea; in any case, it is for sailors the telltale sign that danger is near. In sum, therefore, we can say that the category of *vox confusa* comprises sounds that are usually expressive or communicative and always significant. Since this is true a fortiori of the *vox articulata*, we would seem to have arrived at a new possible solution for the translation of the grammarians' *vox*: namely, "significant sound." Such a phrase rings a bell well beyond the Stoic and Varronian tradition, since Aristotle, as we saw in Chapter 1, influentially defines language (*logos*) as *phōnē sēmantikē*, which we might render as "significant sound," except that he takes care to clarify that he really means the "signifying voice" produced by a human subject; this can be captured by the letters of the alphabet and must therefore be distinguished from *agrammatoi psophoi*, "noises that cannot be written down," including those made by animals, which may function as signs but still are not the products of rational beings.[57] We could continue this philosophical tour, the full version of which would require an extended stay in the Garden of Epicurus,[58] but we have seen enough to begin to suspect that *vox* and *phōnē* are just back-formations of language into sound, yielding a voice-language system in which the understanding of the voice serves primarily to legitimize a particular understanding of language. This sleight-of-hand hardly makes the voice less interesting. Like the edge that joins two sides of the same coin, the voice defines both language and sound because it delimits them, not in some final sense but in a way that sends that coin endlessly flipping: the voice is whatever pulls sound toward language, and vice versa.

In a short chapter, Gellius takes up the contested question of whether the *vox* was a physical thing (*corpus*) or not, but he quickly abandons the whole debate as pointless and potentially endless, concluding by quoting Ennius against the whole bloated enterprise of philosophy.[59] In fact, as readers of Gellius, our task is not so much to pinpoint his concept of the voice (he does not have one) as it is to locate what in the experience of art and life he is pointing to via the term. This much seems clear: his matter-of-fact aestheticism hardly presents us with the voice as it is imagined, for example, by the Lacanian critic Mladen Dolar (whom we met in the introduction), that is, as a seemingly prelinguistic or preterlinguistic sonic real we are condemned to desire hopelessly, with nostalgia or neurosis. For sound stays with Gellius, even through his nights of silent (or, at least, solitary) surrender to the symbolic. Latin itself gets the Gellian view right: wherever there are words (*voces*), including written ones, there are voices (*voces*). Words, of course, are something of which antiquity has left no mean supply. Gellius himself knew this, already in the second century CE, looking back on a classical tradition that, though still part of his social life, he largely treats as closed. Both humanist and classicist avant la lettre, Gellius was also a philologist in the most etymological sense of the word, "a lover of words." Before we go, therefore, let's go find him in his study, or perhaps at the bar, for a few final thoughts, if not entirely in a Gellian mode, then perhaps in a Barthesian one, about the words of our song.

As some readers will have recognized, the brief song reworks and miniaturizes one of the most famous episodes in Homer, in which Thetis entreats the god Hephaestus to forge new armor for her mighty but vulnerable son Achilles.[60] And as every classicist trained in the last few decades knows all too well, the most famous component of that armor—its shield, lavishly described in antiquity's most imitated ecphrasis—can be read as an emblem for the *Iliad* as a work of art (like the shield itself), as the product of work (like that of Hephaestus), and as something that must continue to work, through the work of its reader or singer, in order to animate its story (like the figures shown on the shield, which Homer says seem to move, including some that are making wine, to the sweet song of a lyre-playing boy with a "delicate voice"[61]). Accordingly, our Anacreontic miniature is full of the language of artistic facture, from the first line's highly specific verb for delicate metalwork, *toreuein*, to, three times, the far more ordinary *poiein*, the verb used by Greek artisans who signed their work (so-and-so "made" this) and the verbal root of "poetry" itself. To the working of the cup we may add the work it depicts: the "trampling" (*patein*) of the grapes destined to become what the cup has been designed to hold. There is something on display here, however, other than the "work" in all its senses, namely, the matter that is worked on: the silver that gives the poem its opening words, the gold with which it is to be chased near the end, and between them, the grapevines laden with fruit. Of course, each of these is the product of prior work, for silver and gold must be mined and refined, and grapes do not easily grow on their own. Far from diminishing the role of material here, these bring into sharper focus a basic property of work, namely, that it pushes whatever is worked on toward the status of pure matter, still waiting to be shaped. This is true of silversmiths and their silver, of wordsmiths and their words, of singers and their songs—and even of readers and what they read. Not only must we all work, but we all also must work on some*thing*. To recognize this is to begin to appreciate the thingness of the art object in its own right, to put down the pheno-cup, and to raise the geno-cup in its stead. If the former is, at best, half-full for us, the latter arguably still runneth over. The aesthetic spectacle here, whether visual or auditory, is emphatically one of becoming rather than being, and this is enriched rather than impoverished by the fact that neither the complete cup nor the complete song is really there, plain as day, on the page. In what only seems a paradox, a text's very inability to represent an aesthetic object fully is key to its ability to capture—or, at least, to conjure, but is there really a difference?—a good part of what *matters* about such objects.[62]

A corpus of *Anacreontea*, including a version of the song in Gellius, was rediscovered in the Renaissance and published as the work of Anacreon himself; its influence on poets of the ensuing centuries was enormous.[63] Modern classicists tend to regard that reception as doubly misguided: the poems are not really by Anacreon, and they actually are not very good, endlessly recycling trite themes and figures. Skeptics are correct on the first count, but their correction of the taste of poets from Ronsard to Goethe gives pause. Even Martin West, who liked the *Anacreontea* enough to provide them with a critical edition, dares, in the *Oxford Classical Dictionary*, to say only, "There is no depth in them, but they are often delightful."[64] He is right about the former only if we sound no deeper than the shallow *repoussé* of story and sentiment. On such terms, the following is just another song about wine and love, this time in the service of a formulaic *carpe diem*:

On myrtles tender
and leaves of lotus
I want to stretch and raise my drink.
Let Love in a tunic
tied closed with papyrus
serve me sweet wine.
Like the chariot's wheel
our life spins along:
thin layers of dust
when our bones are released.
Why spread myrrh on a stone
or waste wine on the dirt?
Perfume me rather
while I yet live.
Crown me with roses!
Call for a girl!
Before I go down there
to dance with the dead
I want to kiss trouble goodbye.[65]

Consider, however, something deeper: the song's repeated interrogation of matter. In his sixteenth-century imitation of the piece, Pierre de Ronsard zeroes in on this theme, even as he transforms it, substituting grass for the lotus, a shady laurel for a myrtle bed, fabric of linen or hemp for a papyrus knot, and so on.[66] He thereby offers an emblem for the transformative work that any poetic engagement (ecphrastic or not, imitative or innovative, authorial or readerly) necessarily must do with whatever it encounters as its notionally "raw" material. In other words, embodied in the subject "matter" of this Anacreontic poem and its tradition is a complex meditation on poetic materiality itself—"the theme of *la parole et le marbre*," as James I. Porter calls it, borrowing the title phrase of Jesper

Svenbro's study of Greek lyric.[67]

At the simplest level, matter matters *in* art because matter matters *to* art (and to artists): artists make stuff, out of stuff. The persistence of matter in art, through to and beyond the moment of the "finished work," thus pulls us endlessly back from product into process. Artists have long figured this pull as a still deeper gravity, one that attracts them back even to the *rudis indigestaque moles*, the "crude and undifferentiated mass," from which the world itself is said to have been made. The phrase is Ovid's, and he places it at the beginning of what he calls a *carmen perpetuum*, a "continuous song"—or even a "long-playing record." Of course, the *Metamorphoses* is precisely the kind of poem for which the name of "song" strikes our ears as artificial. Perhaps, however, our evening in the Roman countryside has given us a new way to understand why the ancients called all poetry "song" and traced its history back to the *Ur*-songs of Homer. Far from recalling a lost paradise of unmediated art, they are calling us back to matter, through the most insistent materiality of them all: that of the human body. If we take the call of the voice to be that of matter, forever calling us away from the reduction of *voces* to disembodied words, then the world of ancient literature becomes a very noisy party, once again.

TRACK FOUR

Are You Experienced?

> How much did this people have to suffer
> in order that it might become so beautiful!
>
> —FRIEDRICH NIETZSCHE, *THE BIRTH OF TRAGEDY*

The experienced student of Greek tragedy will have caught, in the closing arguments of the last chapter, something of a Nietzschean strain. For in *The Birth of Tragedy from the Spirit of Music* (*Die Geburt der Tragödie aus dem Geiste der Musik*), to give his first book its full 1872 title, Nietzsche argues that the paradigmatically Dionysian force of music opposes a "Primal Unity" to the division and order of the Apollonian *principium individuationis*. Furthermore, many of the conclusions of my Chapter 2 partly reprise Nietzsche's detection of an especially oppressive version of the Apollonian in the endless Socratic search for knowledge, parroted by the "enormous historical need of dissatisfied modern culture" and its own "consuming desire for knowledge."[1] Indeed, the present inquiry inevitably shares with Nietzsche, if not a zeitgeist, then at least a common longing for the *Geist der Musik* and other sources of pleasure and astonishment to which certain kinds of thinking seem definitionally indifferent:

> Anyone who wishes to examine just how closely he is related to the true aesthetic listener, or whether he belongs to the community of Socratic, critical human beings, should ask himself honestly what he feels when he receives the *miracle* [*Wunder*] presented on the stage...[2]

Accordingly, we turn in this chapter to Nietzsche's signature terrain of tragedy. Here, however, we shall part ways with the German philosopher, and not only because ours is not a Wagnerian quest for national myths and native musicalities. We shall begin instead five years after *The Birth of Tragedy* and an ocean away from German *Sturm und Drang*, with the quieter sounds of a nursery rhyme, recited into the machine this book already has several times invoked, beginning with its title. We shall spend quite a while with Edison's invention, retraining our ears before listening for the fainter sounds of its ancient counterpart. We shall then consider the peculiar voice of a most unlikely ancient performer of tragic repertoire: the Roman emperor Nero. We shall wind up not so much with tragedy's birth as with an enduring aspect of its essence, which we shall locate not in Dionysian music but in the "Heraclean" voice.

Manca la diva

We have come to regard the phonograph as an improvement on the text precisely for its unprecedented ability to "write the voice," which, as we already have had occasion to note, is what the word means. Properly speaking, however, the phonograph records not one but two aspects of a vocal event: the specific sounds made by a voice, and the situation that requires those sounds. When the vocal event is an artistic one—*Macbeth*, for example, whether Shakespeare's or Verdi's—it has become customary to designate these two quantities as the "performance" and the "piece" (though more improvisational vocal art forms, like jazz, will of course muddle this distinction). Thus, in the first phonographic recording ever made, we might say that the performance was Edison's, and the piece was "Mary Had a Little Lamb." Edison, it seems, merely recited—that is, he did not sing—the words, and so the piece, as distinct from the performance, arguably resided no more comfortably on his cylinder than it had in books of nursery rhymes. Had he sung, his cylinder would still have been no better a record of the piece *per se* than a sheet of musical notation. What his invention instead offered was a chance to record the performance, i.e., the specific sounds made by a voice—in this case, Edison's, on that particular day, in the particular mood of that most particular moment, and so on.

Nevertheless, it would be a mistake to assume that what the phonograph captured was, therefore, "Edison's voice," plain and simple. Fast-forwarding through the aftermath of Edison's invention, what we mostly find is singing, which soon became the paradigmatic (though not exclusive) vocal event to record. And while writing and musical notation had long before made it possible to write a song before any singer sang it, Edison's invention did not perform an analogous and opposite miracle, "writing the voice" without also recording the song it sang. Professional singers call the vocal apparatus an "instrument" precisely because it is a *potential* for sound; the phonograph could no more record it per se than it could an unplayed violin. The proof that no single recording of, say, Maria Callas singing *Tosca* could capture her voice has long been there in collectors who fill their cupboards (or computers) with her every *Tosca* and all her other roles in all their recorded performances. And even then, the point is hardly synthesis but, rather, an ensemble of specific pleasures: *this* performance of *that* piece. *Vissi d'arte*, etc.: far from separating art *for* the voice from art *of* the voice, the phonograph traces a single groove for both.

So, in 1953, with EMI's tape recorders rolling, Victorien Sardou's original French play, *La Tosca*, in its Italian translation and adaptation by librettists Luigi Illica and Giuseppe Giacosa, set to music by Giacomo Puccini, sung by Maria Callas, conducted by Victor de Sabata, came together in Milan to realize a voice. Whose voice? If we have to give just one name, it must surely be none of these but, rather, that of someone who never existed: Floria Tosca. To regard the other names in this list (including EMI) as, in essence, *media* for the recording and transmission of Tosca's voice is not to suggest that she is their point: on the contrary, the medium is decidedly the message here. But it is as "Tosca" that these media are most easily compared: "Callas's Tosca" versus, for example, that of Leontyne Price. And while the possibility is long lost for us, some lucky theatergoer conceivably existed whose memory enabled him or her to attempt a more complex comparison of the former to the Tosca of the so-called *voix d'or*, Sarah Bernhardt, for whom Sardou wrote the role of the original play and who would share with her singing successor the epithet "Divine."

That a "role" has something to do with media is arguably implicit in the term's derivation from the "rolls" from which medieval actors learned their parts. While this might lead us to imagine performers as so many player-pianos, the repeated role (across art forms, across performers, across performances) reveals a more complex mediation: a continuous (re)writing of the role on and through the voice, its transcription as both transmission and transformation. The roll, let us say, is one medium of the role-made-flesh; others, whether books or bodies, may follow; and though the role may appear as the thread that binds them, it, like Tosca herself, has no real existence outside of these impressions of it. Nor can we distinguish Sardou's script, Illica and Giacosa's libretto, Puccini's score, and EMI's album from Callas's voice by supposing that this last, unlike the others, does not really *record* anything, if only because when she again sang the role for the microphone a decade later, her voice showed clear signs of wear, scored, as it were, by the hundreds of Toscas, Normas, and other roles sung in the interim.

This later, damaged voice, in fact, has its own distinct allure. Here is one critic's attempt to describe the difference between the two recordings: "Many things had happened to Maria Callas by 1964. Purely as sound her last Toscas were often unfortunate. Whenever the singer approaches the top of the staff, her voice becomes strident. But as a portrait of Puccini's fullest character, Callas's mature Tosca is a rich achievement, worthy of your deepest attention."[3] The "many things" that open this assessment turn out not to be oversinging or the vocal strain often supposed to have been the result of her deliberate weight-loss in the 1950s but, rather, the vicissitudes of her personal life:

> Tosca's act I duet with Cavaradossi was for the Callas of the early 1950s a study in flirtation. In it she was pert, coy, even frivolous. The diva did

> know the meaning of jealousy in those early years and she brings this to bear on the scene, but she did not know what it meant to be in love. The 1964 duet seethes with sexual passion. In it we hear precisely the mixture of religion and eroticism Puccini intended. It took Callas a long time to discover this insight, but once she did, she portrayed it perfectly.[4]

In other words: practically virgin vinyl in 1953, Callas came to be more pliant. Trite as this fantasy is, it belongs to a more general vocal aesthetic that mixes mind and body under the sign of *experience*. "Are you experienced?" is, of course, a question we associate not with opera but with such Callas contemporaries as Jimi Hendrix, Bob Dylan, and Billie Holiday. Callas, however, can usefully be compared to this last, whose final studio recordings were long neglected in the belief that her voice, ravaged by drugs and drink, was too far gone at that point to be pleasurable. Today, these same recordings are regarded as some of Holiday's most signature—deep-grooved versions of the agonizing beauty that had been hers all along. So, too, do Callas's later slips emphasize the perils of the vertiginous edge that had distinguished her voice from the get-go, on which point it is worth remembering that the now celebrated 1953 recording was faulted, in its day, by critics who preferred the cleaner sounds of Renata Tebaldi, to whom Callas's detractors would continue to compare her unfavorably throughout her career.[5] Mystified by her critics to the end, Callas herself would eventually look back on her later, edgier singing as simply the logical culmination of what had always been her own particular art.[6]

De gustibus nil est disputandum. The moral of this operatic interlude, however, is more than this. To bring a more important lesson into focus, we must take a quick trip to the future. Let us imagine an archaeologist millennia hence, long after we and our music have been forgotten, who unearths a mostly intact set of the 1953 long-playing vinyl records. Divining their function and constructing a crude device to play them, he listens to the strange ancient sounds they preserve. He has no ear for their beauty—almost everything about musical experience is culturally specific and thus must be learned—but he recognizes amidst the din the sound of human voices, speaking Italian, a long-dead language that has only partially been deciphered by his colleagues. Just one of these voices is female, and she seems to enjoy some special importance in the group, which calls her "Tosca"—the same name prominently written, in an ancient alphabet, at the center of each disk. The name is familiar to our archaeologist, and he soon connects it to other artifacts bearing the same inscription and belonging, within the margin of error of the tests he can perform on them, roughly to the same era: a text (in which we would recognize Illica and Giacosa's libretto) that minutely records the same words pronounced on the disk by Tosca and the rest; another text (Puccini's score) that adds an indecipherable notational system that presumably records their music; another still (Sardou's play) that seems to be a translation of the first into another ancient language, French. Meanwhile, his excavation yields the 1964 recording, which shares with the other not only the large-lettered "Tosca" but the smaller subscription "Callas." Then another set, but marked "Tosca: Tebaldi." Our archaeologist publishes his findings with this hypothesis: Tosca was the title of a singing Italian priestess (and perhaps also the name of the goddess who inspired her), served by a college of likewise singing priests. Known priestesses were Callas, who apparently served for an especially long time, and another named Tebaldi. They sang a prophetic hymn, carefully learned from their predecessors. Each of the Tosca-inscribed artifacts was an effort to record and disseminate these sung prophecies, which eventually found an audience in France. The archaeologist is soon rewarded with a prestigious chair at a major university.

This future bungler, I would like to suggest, falls victim not only to time's deletions but also to two time warps that already lie, mostly invisible to us, in our various *Tosca*s. The first warp is that of tradition. Within a tradition, in the absence of information that properly belongs outside that tradition, although it may travel alongside it, cause and effect become indistinguishable. If we do not at least know that Greece comes mostly before Rome, and both before England, then we cannot easily tell whether Homer derives from Virgil or vice versa—or whether both derive from Chapman. Even for scholars of a tradition whose ostensible primary goal was exact duplication, as was the case in the medieval copying of texts, error is a degeneration whose directionality, absent external criteria, must be hypothesized and defended on the basis of principles (e.g., that of the *lectio difficilior*) that offer likelihoods rather than certainties. And such principles are of little use in sorting out a tradition of creative imitation. Indeed, nothing but history (and its various date stamps) prevents us from supposing that Sardou imitated and adapted Puccini, rather than the other way around.

The second time warp is that of performance and its representations. Armed with pen and paper, a member of Sardou's audience for the play's initial run who wrote down the players' words could have produced something similar to the script from which they had learned their parts, especially if she or he also recorded the movements (*exits, stage left*, and the like) and tones of voice for which Sardou had provided extraordinarily minute instructions. Likewise, a musician with a copy of a recorded performance of Puccini's opera could, with the aid of repeated and meticulous listening, generate a complete score, generally differing from the composer's own only to the extent to which it had been modified for performance by the conductor. Scripts and scores, in other words, are usually taken to be *prescriptive* because we mostly use them this way, but nothing about the objects themselves would distinguish them from something instead made, during and after the fact, for *descriptive* reasons. In the same way, those representations of performance we normally would call descriptive, such as record albums, can be deployed prescriptively: Tebaldi to judge Callas, or both to judge a new singer, condemned to compete with all her recorded predecessors. Nor need these uses follow any straightforward timeline: Callas 1964 can train us to hear what is "missing" in Callas 1953; either or both may affect how one reads Puccini's score—or even Sardou's play. One can "learn the part" of Tosca from any performance or representation, prescriptive or descriptive, thereof, and it is only as these have unfolded in time (and have been registered in and as history) that they may be said to belong to a particular sequence of artifacts and events.

Indeed, even in the liveliest moments of a performative tradition that is playing out in historical time, these same constitutive time-bending phenomena, tradition and performance, function to destabilize what would seem to be the two most basic of artistic quantities: the artist and the work. Consider the title page shown in figure 7, produced even as Puccini's opera was enjoying its premiers in Italy and abroad. The stern notice at its center admonishes, "All rights ... are reserved." But reserved to whom? That is, whose *Tosca* is this? Possible contenders blanket the page: Sardou, Illica and Giacosa, English translator W. Beatty-Kingston, Puccini, G. Ricordi (and Company). Even the book's various collectors get into the act, starting with B. N. Freeman, whose name is handwritten on the title page along with a date of purchase. Subsequent owners are recorded on the bookplate: Edwin Corle, who had married Freeman's daughter (the actress Helen Freeman) in 1932; Jean Corle, Edwin's second wife and eventual widow; UCLA, to which she donated the book—and from which this volume once claimed by so many has now been sent to lonely offsite storage at the Southern Regional Library Facility, named on a sticker partially visible at the corner.

Owner's marks provide one kind of answer to the question, "Whose *Tosca* is this?" The copyright notice at the bottom of the title page offers another (albeit one likewise able to be superseded over time: in copyright terms, this book is now no one's and everyone's). The librarian, however, must find here the name of an author. For this book, the catalogue settles on Puccini, a seemingly reasonable choice, since his name is largest and the warning at the center refers to the rights to an "opera." However, what follows the title page is not, in fact, Puccini's score, but rather a plot summary, followed by Illica and Giacosa's Italian libretto, followed in turn by its English translation. In other words, this book was intended primarily for sale at the opera house, to patrons who would have taken their seats clutching it along with a playbill that offered still other possible answers to the same question. At the opera's New York premier on

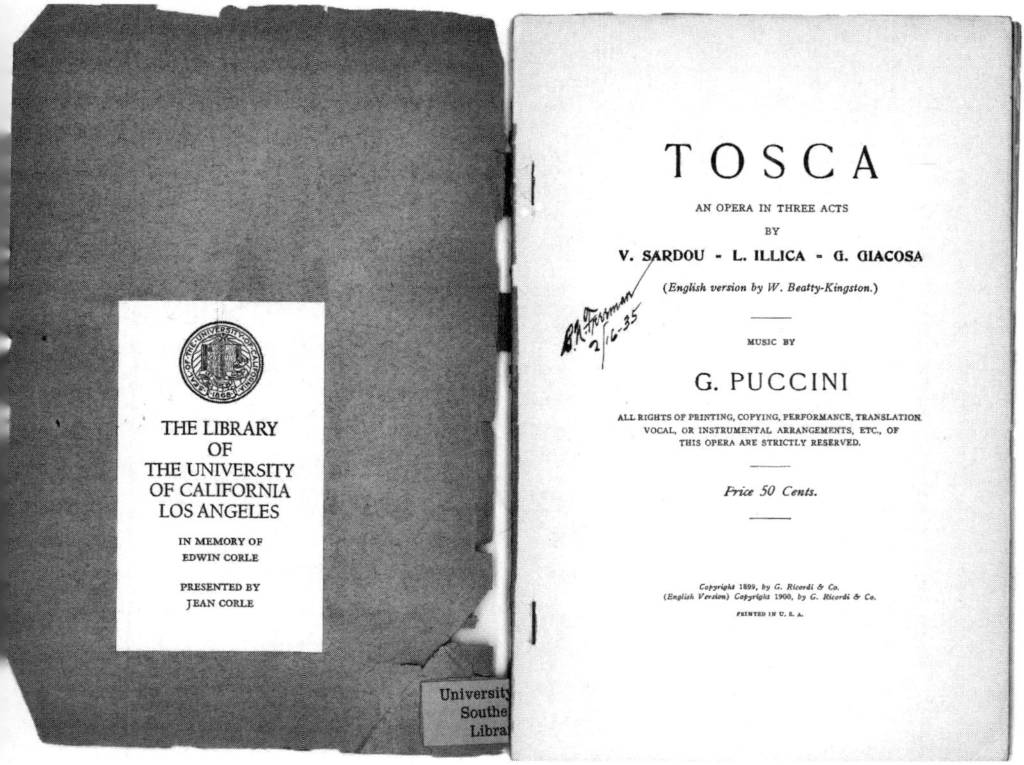

Figure 7. *Tosca*, from the Library of the University of California, Los Angeles.

February 4, 1901, these alternatives would have included conductor Luigi Mancinelli and, of course, that night's Tosca, the once famed Croatian soprano Milka Ternina.

A few days later, in ominous anticipation of Toscas to come, Ternina was described by the opera's reviewer for *The New York Times* as a singer who "has developed dramatically at some sacrifice to her former perfection of singing."[7] In fact, assessments like this one, and the implied contest of taste that lies behind it, can be paralleled throughout the performance (and recording) history not only of *Tosca* but of opera generally. I would like to suggest that these critical mantras add up to something rather more interesting than the sometimes conflicted dual nature ("musical" versus "theatrical") of opera itself. The question usually presents itself to us as that of the most effective mode of operatic *expression*. But by a kind of Newtonian law of our mediatic imaginary (traceable back to the Aristotelian principles explored in the introduction), we cannot help supposing that to every expression corresponds an equal but opposite *impression*. In other words, we tend to treat the expressive voice as an impressible *medium*—one that through expression makes an impression in turn on us.

Our disagreements fundamentally regard the imagined sources of that impression on and of the voice. In a nutshell, one can sing *Tosca*, or one can sing Tosca. The first kind of singing seems to us to receive, faithfully, the imprint of the role's various authors. The other kind noticeably modifies and supplements those scripts. This latter can be characterized, by fan and foe alike, as the assumption of an authorial role by the singer herself, who leaves her own additional impressions on the part. Those who favor such singing, however, tend simultaneously to imagine it as the result not so much of heightened agency as of greater susceptibility on the part of the singer. Susceptibility to what? By the same Newtonian law, each authorial expression seems to depend on an impression. Within a tradition, this impression notionally comes from one's predecessors: Illica and Giacosa express what they receive from Sardou, and so on. In another sense, however, all of *Tosca*'s authors record impressions of the things their tradition will come, retrospectively, to embody. Chief among these "things" is, of course, Tosca herself, whose name looms larger than any others on this and other title pages, playbills, and marquees. Once again, whose *Tosca* is this? The best answer, I would suggest, is always that it is Tosca's, not because the tradition convinces anyone that she actually exists, but because the illusion of her existence is itself more useful than any of her single authors/ artists (whether playwright, actor, translator, composer, or singer) to explain what these multiplying *Tosca*s have in common. Whatever else they may do, they all record *her*. Indeed, deep into a tradition, it becomes difficult to imagine even its original inspiration as anything other than an impression of what, once it came, was always already there.

Of course, in a sense, Sardou's Tosca really was already there, since he wrote the part for a voice that had already left its deep impression on him (and the world), namely, Bernhardt's. In fact, to the extent to which his script presupposes her expressiveness, it can be said to offer a precious early record of the *voix d'or*. And while the decision to make his tragic heroine an operatic diva seems to offer a prescient nod at *Tosca*s (and Toscas) to come, it was first of all a sidelong wink at the voice of his "divine" actress. In 1902, fifteen years after she first played *La Tosca* and a year before the Paris premier of Puccini's opera—for which, in a twist of fate, Sardou himself would provide stage direction—Bernhardt's voice would first be recorded by the phonograph, in a performance of Racine's *Phèdre*. Here too, writing the voice (the phonograph), writing for the voice (Racine), and vocalizing (Bernhardt) all give shape to Phèdre (and *Phèdre*), reminding us that Tosca's is hardly the only tradition in which distinguishing between creating a voice and capturing one seems to miss the (phonograph's) point.

Of course, one might suppose that it is hardly a coincidence that the phonograph has emerged as paradigmatic in an analysis I began, in the introduction, with Edison. Irresistible for us to think with, the phonograph was, already by the end of the nineteenth century, as Friedrich Kittler has taught us, impossible to ignore;[8]

Sardou, Bernhardt, Puccini could hardly have escaped its influence. Nevertheless, proof that some of our age's phonographic instincts predate Edison can be found in their anticipation, centuries before, by Racine—and not only because he too wrote (for) a famous voice, that of Marie Champmeslé, his first Phèdre. The first relevant detail is something too obvious ever to receive notice: Racine chose the name of his play from those of his dramatis personae. In this case, his original title was actually *Phèdre et Hippolyte*; he thus followed both of his ancient models, Euripides' *Hippolytus* and Seneca's *Phaedra*. The convention is in fact a venerable one: if we include names for groups of characters (e.g., *The Trojan Women*), then not one Greek tragedy can be said to deviate from it. One can then distinguish between title and title role by prefixing to the former, "the tragedy of," or by applying to it modern typographical conventions like inverted commas or italic type, but throughout history, the difference mostly has gone unmarked. This sets up the same homophony to which I have already called our attention: to hear *Antigone* is to hear Antigone, and vice versa—simply put, they sound the same, at least while the title character is speaking. Of course, sometimes one must be more specific: to hear Euripides' *Medea* is to hear Euripides' Medea, but this is not exactly the same as Seneca's, not least because the latter speaks Latin. Sometimes, in fact, a shift in language is reflected in the shared name: *Phaedra* becomes *Phèdre*; *Heracles*, *Hercules*. Naturally, by giving us the names of protagonists famous in myth and literature, these titles also tell us more or less what the play will be about. This, however, is a convenient byproduct of the more basic mechanism by which they tell us not *what* we are about to hear, but *whom*, and even, to some extent, what they will sound like. To go to Racine's *Phèdre et Hippolyte* is to know that you will hear Phaedra and Hippolytus, in French—and not just in any French, but in Racine's.

The importance of this seemingly trivial observation will be clearer in a moment. First, however, we must turn from the conventions of tragic titles to those of tragic settings. Racine's play is, of course, set in the past—indeed, in antiquity itself. So too had

Euripides set his own *Hippolytus* in an age that was already ancient for his contemporary Athenians. Indeed, whereas Greek comedy prefers the present and may even venture into the future, Greek tragedy is always set in the past.[9] There are, of course, many reasons for this, but the most important is regularly overlooked: a setting in a more or less distant past tells us that, if the tragedy's characters ever existed, they are by now most assuredly dead, violently or not. Whether or not the audience already knows the plot of the play, it knows from the outset that Phaedra, for instance, is no more, long before she hangs herself in Euripides, or stabs herself in Racine. To take this small fact seriously, I would suggest, is to recognize that tragedy attracts its audience not so much with living voices as with reanimated ones. (Compare opera: meditating on the long afterlife of Orpheus, including that of his severed head, Carolyn Abbate observes that "opera itself could be regarded as response to an outlandish question—how does the dead object continue to sing?")[10] Indeed, we may go further still, and say that the tragic voice itself, abstracted from any particular character, is not a living voice but a perennially undead one: kill it off at the end of one play, and it will always come back in the next.[11]

Let us return to that most modern and mortuary participant in a vocal tradition: the record collector, whom we again imagine in a nearly extinct guise, lining his shelves with paper-shrouded vinyl in cardboard coffins. "Similar to the fate that Proust ascribed to paintings in museums, these recordings awaken to a second life in the wondrous dialogue with the lonely and perceptive listeners, hibernating for purposes unknown," as Adorno puts it in "Opera and the Long-Playing Record," a final return to the questions of phonographic writing that had exercised him in his youth.[12] The collector may regret not having heard this or that piece or performer "live" or may mourn the records that he cannot have because they were never made, but for the most part, death is no more serious an obstacle to his listening pleasure than is the fact that Phaedra and Tosca never really existed. He has his critics, of course, most notably the managers of theaters and opera houses who would prefer he spend his time and money on their "revivals" of the classics, or on pieces newly born. In fact, he may well be a ticket buyer, or even a season subscriber. But back home, he repeatedly demonstrates the uncanny comfort of these performative traditions with death. To suppose that this is because he fails to appreciate the difference—"is it live, or is it Memorex?" as an advertisement for another dead technology once asked—is to ignore the fact that Phaedra and Tosca have never presented themselves as anything *but* records, i.e., as ostensibly remembered voices from the dead and distant past. These have been made of paper and ink, costumed flesh and blood, vinyl: not just mediation after mediation, but (further) mediations *of* mediations. To the suggestion that he is missing something, the stay-at-home record collector can quite reasonably respond that, on the contrary, no one knows a truer sound of the tragic voice than he. Even his medium's inevitable scratches and hiss only amplify the results of each tragic hero or heroine's fatal flaw. After all, didn't Aristotle say that truest tragedy was about bad things happening to relatively good vinyl?[13]

From *Ajax* to *Aida*, the tragic plot has enjoyed a remarkably long (if not entirely continuous) run. Our record collector, I would suggest, may be the best person to help us finally to understand why. Of course, this question has received many answers over the centuries, and with a little effort, our collector can be made to correspond to aspects of these, demonstrating, e.g., edification through empathy at a safe distance (Aristotle) or the tension between Apollonian individualism and Dionysian self-abnegation (Nietzsche). But he ultimately points us toward something far simpler. For whenever our collector talks about his records, he is forever talking about voices. While we might be inclined to attribute this fact to the operatic and phonographic character of latter chapters of the tragic tradition, these arguably only amplify what was present from the beginning—indeed, there just beneath the nose of every tragic mask. To be clear, I do not mean merely to say that tragedy exploits the voice. I mean rather that it is possible to imagine tragedy as a *consequence* of the voice, even at the level of plot.

This doubtless seems an ambitious claim, and before I attempt to defend it, let us spend a final moment with our collector. Gladly listening to operas in languages he does not understand and still growling bitterly about the day that opera houses began projecting supertitled translations, he poses by his very existence a direct challenge to any assumption that tragedy is fundamentally a *verbal* art. Even when he knows the story, he delights in being taken to places where words fail, even if they do not entirely stop. (If you really want to understand Callas's two *Tosca*s, compare her screams when she discovers Cavaradossi is dead, absent in the libretto and scored simply as an unsung shout: "Ah!") The tragic voice expresses itself to him *in* language but not necessarily *as* language; plots that maximize that dissonance keep the voice center stage—or at least, dead center between his stereophonic speakers.

Whether or not he has ever seen or heard an *ancient* tragedy, our collector offers a lesson to scholars of classical Athens. Especially in the wake of the work of Jean-Pierre Vernant, we have tended to assume that what drives tragedy is a fundamentally *linguistic* crisis, attributable to broader tension in the semantic web of Athenian political life.[14] What happens, however, if we reverse this relationship of (sociolinguistic) cause and (artistic) effect? In other words, could it not instead be the case that tragic art went looking for linguistic trouble, something the Athenian polis—hardly unique in this regard—was all too ready to provide? To explore this further, let us turn off the opera and turn in earnest to antiquity, not, however, to the democratic frisson of classical Athens, but to the no less noisy tragic stage of despotic Rome.

Stormy Weather

Whether or not he really fiddled while Rome burned,[15] Nero was something of a musical sensation, singing to packed houses throughout his empire. His success, the unimpressed historian Suetonius reports, came in spite of the fact that his voice was "small" (*exiguua*) and "dark" (*fusca*).[16] The second of these is ordinarily used of "swarthy" human complexions; thus Gallus, in Virgil's *Eclogues*, defending a singing shepherd boy he loves, cries, "So what if Amyntas is *fuscus*? Violets and berries are dark (*nigrae*) too!"[17] Nero, though, was not himself *fuscus*: Suetonius later reports that his body was instead *maculosum*, by which he presumably means it was pale and "splotchy."[18] Nor was that body exiguous, he goes on to say, with the exception of its spindly (*graciles*) legs: the emperor was average height (*statura prope iusta*), had a potbelly (*ventre proiecto*) and, significantly, a giant, jowly neck (*cervice obesa*), a feature plain to see on coins minted with his portrait. Into that operatic throat, Nero would finally drive a dagger, with the help—which one can well imagine him needing—of a former slave named Epaphroditus, "Lovely."[19]

Nero's voice need not have been as bad as Suetonius says, nor even bad at all. This scandal-sheet biographer is anxious to draw as much contrast as possible between any actual talent Nero may or may not have had (Suetonius himself, born around the time of Nero's death, could never have heard him) and the Emperor's international success as a performer (the Roman crowd cried out for his *vox caelestis*, "heavenly voice," and he won first prize in singing competitions throughout Greece), which Suetonius again and again attributes to the desperate sycophancy of a world with little choice.[20] Whether Suetonius's complaint is fair, however, it must at least be a plausible thing to say about unpleasant singing. While it is clear enough why a "small" voice is a problem, what exactly is faulted by the synesthetic metaphor of a "dark" one? Both terms would seem find an echo in the Greek satire *Nero*, of uncertain authorship and date (though probably long after Nero's lifetime), which pillories the emperor's vocalizing as "naturally hollow (*koilon*) and low (*baru*)," adding that the strange anatomy of his throat means that "his singing makes a kind of buzzing sound."[21] And a story in Plutarch that has Nero reincarnated as a singing frog (a possible source, along with apocalyptic literature, of an odd medieval tradition that the emperor once gave birth, through his mouth, to the same animal) may add a further epithet, "froggy," to the surviving lexicon of Nero's venomous critics.[22] All of this, however, only adds insult to injury. Can't we find anything nice to say?

We are helped by a remarkable list of antonyms in another text. The ear, Cicero tells us in his *On the Nature of the Gods*, can distinguish between "very many different vocal kinds" (*vocis genera permulta*), including the smooth (*leve*) and the rough (*asperum*), the deep (*grave*) and the high (*acutum*), the bendable (*flexible*) and the stiff (*durum*), and—this is actually the first pair in his list—the sonorous (*canorum*) and the dark (*fuscum*).[23] Since dark's opposite here is just the adjectival counterpart of the Latin words for singing (*canere*) and song (*carmen*), all we really learn is that a dark voice is an unmusical one—and this already was implicit in Suetonius's criticism. We have, however, somewhat raised the stakes of our terminological inquiry, for "darkness" now emerges not merely as a musical fault but as something like an opposite of music itself.[24] We are reminded, perhaps, that the Greeks and Romans gave dominion over both music and light to the same god, Apollo. Olympus itself, Apuleius tells us, was a place forever free from *fuscitas*, "cloudiness."[25] (Sunless Hades, by contrast, is in Aeschylus a place "untouched by Apollo's foot."[26]) By calling Nero's voice *fusca*—unmusical and all too terrestrial—Suetonius turns the claim that it was instead "heavenly" doubly on its head.

Not just Apollo but Venus too is opposed to a *vox fusca*, albeit in a very different way. By having sex (*venus*) now and then, Pliny the Elder advises, "athletes who have lost their energy regain it; the voice is restored, when it has degenerated from clear (*candida*) to dark (*fusca*); so too is relief provided for their muscle-pain, for dull vision, and for those who are mentally disturbed or depressed (*mente captis ac melancholicis*)."[27] A singer's distressed voice is a problem of the first order; the athlete's voice here, by contrast, presumably comes under scrutiny as a symptom of a wider malady that affects his general physical performance. It is striking, at any rate, to see darkness linked here to madness and melancholy, more famous as afflictions not of athletes but of poets—and of the occasional Roman Emperor. "Why is it that all those who have become eminent in philosophy or politics or poetry or the arts are clearly of an atrabilious temperament, and some of them to such an extent as to be affected by diseases caused by black bile, as is said to have happened to

Heracles among the heroes?"[28] Thus opens, famously, the thirtieth of the *Problems* attributed to Aristotle. There and in many discussions since, melancholy is the black bile that produces the black moods from which true genius can never really be separated. A melancholic athlete is no good, but melancholy is at least potentially a virtue in an artist. Could this extend to singing? To borrow our own relatively recent color metaphor for music that comes from the depths of anguish and despair, might not a *vox fusca* be just the ticket for a singer who wanted to sing the blues?

A century after Suetonius, the historian Dio Cassius likewise calls Nero's voice "small (*brachu*) and dark (*melan*)."[29] Since he may be dependent on Suetonius or a source common to them both, Dio offers no fully independent confirmation of the characterization, but his Greek does provide a somewhat revealing escalation.[30] Properly speaking, *fuscus* should not be *melas* in Greek, the regular word for "black," corresponding to the Latin *ater*, but, rather, something like *phaios*, neither white nor black but something in between, that is, "gray," like a cloudy sky, or like the dusk, the English word to which *fuscus* may have an Indo-European link. One might, in fact, have been tempted to take Suetonius's *fusca* to mean "gray" in the sense of "indistinct." This, however, clearly is not what Dio imagined. "Gray" or "black," the problem with Nero's voice was not blandness, but darkness, and he represents that vocal quality with the color of melancholy.

Shades of gray, shades of blue? When smoke got in a Roman's eyes, these last might be *caesii* ("blue-gray," like Minerva's); the offending ember had its own color: *cinereus* ("ashen") or the slightly whiter *cineraceus*. Latin offers, in fact, a remarkable palette of grays, from *plumbeus* ("lead-gray") to *murinus* ("mousy") to *lividus* ("purplish-gray," like a bruise) to *canus* ("hoary," the color produced by white hairs or fuzz against a colored background, like a human head, or a yellow-green quince). *Fuscus*, as we already have seen, primarily designates a swarthy complexion. (Thus, in modern botanical Latin, it is not a gray but a brown.) Compare the enigmatic *aquilus*, apparently "dark like water," used by Suetonius to describe the skin of Augustus, who was somewhere between this color and "white" (*candidus*). So too, famously, *niger* sometimes means "black" like *ater*, but when used of complexions, it often is interchangeable with *fuscus*, and in nature, it can describe a wide array of relatively dark things, like the violets and berries of the lament of Gallus. "Gray," for the ancient Romans, in other words, was no simple black-and-white matter, any more than human skin is. (For "pure gray," in fact, botanists must conscript the Medieval Latin *griseus*, a Latinization of the vernacular *gris*, which, like our own *gray*, comes from a Germanic root.)[31] In sum, a Roman could not call a voice "gray," by any of classical Latin's adjectives, in order to suggest that it was "colorless."

The Greeks had recourse to the more neutral gray already mentioned, *phaios*. This is the word used by Plato in the *Republic* to describe those benighted souls who know the black night of pain but not the white light of truth; what they (i.e., most of us) take for pleasure is really only a shade of "gray."[32] (Here too, though, "gray" does not equal "bland": for all of their philosophical dimness, these dark lives, shaded everywhere by suffering, hardly seem to lack intensity.) *Phaios* is also the word given by the *Thesaurus Linguae Latinae* as the Greek equivalent of *fuscus* used of sound,[33] and Liddell and Scott assign to this use the meaning of "harsh," though this hardly seems confirmed by their four examples (among which we find our old friend Julius Pollux), which mostly limit themselves to listing *phaios* as an adjective sometimes applied to sound, along with "black" (*melas*) and "white" (*leukos*).[34] In his classic *Greek Metaphor*, W. Bedell Stanford launches a brief but provocative investigation "On Synaesthesia or Intersensal Metaphor" by considering the expression "white voice" (*leukē phōnē*), given by Demetrius as an example of a curious metaphor that has become familiar by constant usage.[35] For Stanford, such expressions can never be deciphered or defined, because they serve precisely to assert "the oneness and interdependence of words and things against the sterilizing analyses of the scientists."[36] We might take this one step further and suggest that, rather than asking what this or that color means when applied to a voice, we should consider how the very notion of vocal color (up through operatic *coloratura* and beyond) participates in the voice's own inability to be confined to the speech to which it lends itself. Black, gray, or even white, there is more to such voices than bodiless words.

Our attempt here to rehabilitate vocal darkness actually replays a move made by the rhetorical teacher and theorist Quintilian, whose life overlapped with both that of Nero and that of Suetonius. The application of *fusca* to the voice occurs twice in *The Orator's Education*, both times in a long chapter on "delivery" (*pronuntiatio*, which Quintilian uses as a synonym for *actio*, although, he explains, the term in origin emphasizes the voice over the other element of delivery, namely, movement[37]). Quintilian first includes such a quality in his own brief list of vocal antonyms, opposing it, seemingly as a negative term, to the *vox candida*, the Latin version of the "white voice," that is, a "clear" or "clean" one, in the usual translation.[38] But darkness later returns with some ambivalence. In a brief parenthesis, Quintilian notes that the great orator Marcus Antonius had a *vox fusca*, citing Cicero, who heard Antonius and describes his delivery in his dialogue *Brutus: On Famous Orators*.[39] Quintilian, though, has slightly misremembered his source: Cicero calls Antonius's voice instead *subrauca*, "fairly hoarse or harsh or husky," adding that it was so *natura* "by nature."[40] (By contrast, the later rhetorician Porcius Latro, who stayed up late nights working and refused to bother with vocal exercises and other precautions, had a voice that had been "darkened" [*infuscata*] not by nature but by "neglect" [*negligentia*], according to his friend and fan Seneca the Elder.)[41] Quintilian's misquotation offers more lucky evidence for the meaning of *fusca*, but we should not treat the two as exact synonyms. There is not much you can do about a voice that is *subrauca*; so too for a *fusca* one in Quintilian's first passage. In the second passage, however, Quintilian seems to be playing on an ambiguity in *fusca* (but not, probably, *subrauca*), reversing himself entirely to describe such a voice not only as not always a handicap but even as something a speaker might deliberately mimic: "Here a *vox fusca*, of the sort which Cicero says

Antonius had, works magnificently; indeed, there is something in it which merits our imitation."

Cicero had taken pains to explain how Antonius's *vox subrauca* was effective:

> Uniquely in his case, though, this defect was transformed into an asset. For in rhetorical laments, he had a certain tear-jerking quality (*flebile quiddam*), something which was well suited for both inspiring belief and producing pity, to such an extent that he offered confirmation of what is told of Demosthenes, who, asked what the most important thing was in oratory, replied, "Delivery," and then asked what the second most was, gave the same answer, and then the third, the same again.[42]

Flebile quiddam sounds familiar, for it is close to the *flebile nescioquid* which Ovid uses to describe the last sounds produced by the lyre and severed head of Orpheus as they floated down the Hebrus River, as we saw in Chapter 2. Virgil, in his fourth *Georgic*, Ovid's model here, had limited these sounds to a single name, emitted by the poet's head and echoed by the passing banks: "Eurydice!" How to beat Virgil's simple, direct, devastating pathos? Ovid starts by stripping the decapitated poet of words. The lyre, which cannot "say" anything, sets the tone instead, and the tongue follows suit, content merely to "murmur":

> **Flebile** nescioquid queritur lyra, **flebile** lingua
> **murmurat** exanimis; respondent **flebile** ripae.

> Something weepy (but what?) is the complaint of the lyre,
> something weepy is the murmur of his lifeless tongue,
> something weepy is the answer of the banks.[43]

Ovid borrows the resounding banks from Virgil but brings their echo in closer. Here, lyre and tongue already echo one another, and the latter, with its *mur-mur*, even echoes itself. Naturally, *flebile*... *flebile*... *flebile* is not the echo itself (i.e., Orpheus did not say "*flebile*") but, rather, an attempt, like *murmur*, to represent something either less or more than language itself. We cannot know, from

Ovid, what Orpheus said, if in fact he was trying to "say" anything at all, but we do know something about what it sounded like: it was the kind of sound that makes us cry. What, finally, *flebile quiddam* and *flebile nescioquid* have in common is the fact that they take a double distance from language: I myself (Cicero, Ovid) cannot really name (*quiddam, nescioquid*: "a certain something or other") that which he himself (Antonius, Orpheus) expressed not through words but, rather, through sound.

Back to our headliner. Thus far, I have been imagining Nero as a dusky-voiced torch-song singer: the emperor takes the stage to offer us his unforgettable rendition of "Stormy Weather"—which in Latin, as we have seen, would be *Fuscitas*. At least in terms of repertoire, however, we can do much better, for ancient sources give us the titles of a number of pieces he sang.[44] According to Suetonius, for example, it was during a performance of a *Hercules Driven Mad* that a young military recruit, mistaking the chains of the emperor's costume as signs of a coup d'état in progress, leapt to his defense.[45] Let us take advantage of the momentary confusion produced by this misplaced act of valor to slip into an empty seat in the back row. The emperor returns to the stage. *Silentium!* He steps forward and begins again:

> Quis hic locus, quae regio, quae mundi plaga?
> ubi sum? sub ortu solis, an sub cardine
> glacialis Ursae? numquid Hesperii maris
> extrema tellus hunc dat Oceano modum?
> quas trahimus auras? quod solum fesso subest?
> certe redimus: unde prostata domo
> video cruenta corpora? an nondum exuit
> simulacra mens inferna? post reditus quoque
> oberrat oculis turba ferialis meis?
> pudet fateri: paveo, nescioquid mihi,
> nescioquid animus grande praesagit malum.

> What place is this, what land, what worldly spot?
> *Where am I?* Where sun first greets the earth?
> Or north, beneath the frigid star? Or there,
> beyond the western sea, where land gives way
> to Ocean deep? I breathe what air? What ground
> sustains my tired feet? I must be home—
> and yet what makes me see upon my porch
> dead bodies black with gore? Perhaps my mind
> has not yet shaken off the ghostly sights
> of Hell? But must that demon hoard my eyes
> distract beyond the dark? I dare not say . . . ,
> I dread . . . , I know not what my mind foretells,
> I know not what, but large it looms—and foul.[46]

These are not, however, the words Nero sang. In the first place, he almost certainly performed in Greek rather than Latin—especially when on tour in Greece, but probably everywhere.[47] Second, the play from which I have borrowed these lines, Seneca's *Hercules Driven Mad*, like the rest of Senecan tragedy, may not have been written for the stage; performance, it has been suggested, might have been limited to a kind of dramatic reading, and it is not certain that this would have included any music.[48] In any case, and third, these particular lines come not from the play's choral odes, in meters that at least theoretically could be sung, but from lines written in the iambic trimeter reserved for spoken dialogue (here practically a monologue).[49] To these may be compared the iambic pentameter of my translation, the English meter which, by alternating unstressed and stressed syllables, likewise imitates (while regularizing and aestheticizing) the rhythms of ordinary speech.

We are still too far away to hear Nero singing, but these lines are not exactly voiceless. At the most basic level, of course, they offer sequences of letters capable of suggesting the phonemes of speech, pronounced aloud or not. Add to this their meter, unsung but not unvoiced, product of those same sequences and of vowel lengths known to speakers but not always notated by the letters themselves.

More complex is the series of questions that occupies all but the last two lines. Interrogative words (*quis*, *ubi*, *an*, etc.) and word order signal most but not all of these (the question mark being a medieval invention that would not have been present in an ancient copy of the text) without exactly mapping the conventions of intonation by which the speakers of a language distinguish statements from questions, and different types of questions from one another. Finally, the fact that these lines belong to a speaking *persona* in what at least pretends to be a play expressly invites their vocalization, or at least the conscious imagination of such.

There is a still more interesting way in which Hercules' part here calls for a voice—and a talented one at that. If we miss the earlier cues, we can hardly ignore the repeated *nescioquid* at the end. For if we take Hercules simply at his word (this one, but also the questions that lead to it), we reduce the whole speech to a stale exercise in dramatic irony: he does not know that, driven mad by jealous Juno, he has just slaughtered his wife and children, but *we* certainly do. The point of this key moment in the play is instead the opposite: Hercules is slowly coming to his senses; madness gradually yields not so much to ignorance as to disbelief; *nescioquid* hardly means that he knows nothing; deep down, he already suspects the unspeakable (*pudet fateri*) truth. It is his voice, rather than his words, that must express what he cannot (yet) say.[50]

For a number of reasons, therefore, we can say that this text *requires* a voice in order to work. But let us lean a little on the *re-* of that word's Latin root to ask whether this text not only needs a voice but needs it *again*—that is to say, "goes looking" (*requirere*) for a voice it once had. What exactly was that voice, where did it go, and most importantly, if it can be *re*activated, can we not say that this text *re*cords it (from *recordare*, "to call back to mind")? There would seem to be a basic problem here: *Whose* voice are we talking about? Four possible answers present themselves: the voice of the author, the voice of the text (here, the voice of the speaking character, Hercules), the voice of an actor, or the voice of a reader. For the moment, let us follow the hunch that Seneca's plays were not acted but "read" (perhaps, however, by professional performers) and combine these last two categories. We are left with a multiple choice all too familiar to students of literary criticism: (a) the author, (b) the text, (c) the reader. In other words, the question of where literary voice resides prompts the same trinity of possible answers as does the question of literary meaning. Regarding the latter, critics in the last century ever more insistently condemned (a) as the fruit of the "intentional fallacy," finally settled on (c) as technically the only correct answer, but largely continued to act as if they believed (b) and even, sometimes (a). What is ultimately most remarkable about our efforts to locate meaning is the slippage between answers against which those efforts are waged. Might not a similar slippage characterize readerly relationships to voice? For example, if readers think they know what Hercules means (even though this is notionally denied by the text's *nescioquids*), might they not also think they know what he sounds like? Might not their own reading of the text be (or at least, seem to them to be) an activation of that sound, just as it is an activation of meaning they believe they share with the text and/or its author?[51]

The Passive Voice

With all of this in mind (and in our ears), let us return to our singing emperor and to what Suetonius and Dio report as the greatest hits of his repertoire, all of it tragic: *The Bacchants*, *Thyestes*, *Alcmaeon*, *Attis*, *Niobe*, *Canace in Labor*, *Orestes the Matricide*, *Oedipus Blinded*, and of course, *Hercules Driven Mad*.[52] Most of these are stories of family catastrophes, and given the crimes of which ancient sources accuse Nero, including incest and matricide, it is tempting to suppose that the historians choose to remember these particular pieces precisely because they fuel the suggestion that there was not much difference between the man and his masks.[53]

But there may be more to hear in this list. For one thing, all of these plots begin with or lead to self-inflicted loss: Alcmaeon murders his own mother; Agave, leader of the Bacchants, dismembers her own son; Orestes loses his father, by his mother's hand, and his mother, by his own; Hercules, his wife and children, also by his own hand; Thyestes, his children, whom he is tricked into devouring; Oedipus, his father by his own hand and his mother-wife to suicide, along with his sight and his kingdom; Attis loses instead the love of his goddess and his manhood, again by his own hand; Niobe through foolish boasting loses her children to the murderous rage of the gods; Canace, pregnant by her brother-lover and sent a sword with which to kill herself by her father, loses everything. But just as remarkable (and hardly unrelated) is the conspicuous presence of madness in this list: Bacchants are famously mad; Alcmaeon and Orestes, each guilty of matricide, were hounded into madness by the Furies; Niobe is driven mad by grief, and Atreus by desire for revenge; the *furor* of Attis would inspire one of Latin's most remarkable (and metrically daring) poems, by Catullus; and the madness of Hercules was a perennially favorite theme.[54]

On the one hand, as stories of self-centered loss, these myths invite expression through the elegiac powers of poetry; on the other, as stories of almost *incalculable* loss and, in particular, of the loss of one's mind, they take us farther, into the realm of the irrational, the ineffable, the unspeakable. Words are both too little and too much here: to tell these stories, one needs something else, a kind of "underground thunder," to borrow a suggestively sonic figure from one of a handful of tiny surviving fragments of Nero's own poetry.[55]

In this regard, Nero's remembered repertoire is pointedly emblematic of the whole tragic genre, whether or not the historians who report it are also winking at his infamous confusion of world and stage. Indeed, let us step back for a moment from Nero in order to ponder, once again, the long run of the tragic plot itself. Regarding that tradition's earliest beginnings, it has been usual (if not universal) to accept, with various provisos, Aristotle's assertion that tragedy evolved from dithyramb, that is, from choral hymns sung in praise of Dionysus.[56] Following on this, it has been common to assume that motifs like violence and madness were present in the tragic tradition from the start as social and psychic attributes of the god himself. These would then have evolved into the "slender plots" that, Aristotle tells us, eventually gave way to the complex masterpieces of the classical Athenian stage. In other words, according to the prevailing view, the true origins of the tragic plot are to be sought in the oldest, deepest meanings of Dionysiac cult. To be sure, Athenians would eventually complain that their theater had come to have "nothing to do with Dionysus," but as classicists who took up the phrase twenty-five years ago took pains to show, this only means that what once had been about cult had come to speak more pointedly to other aspects of culture (especially the tensions of the body politic).[57] One way or another, tragedy evolved to tell stories. Along the way, it necessarily explored, exploited, and developed the expressive potentialities of the voice.

Or so we have supposed. Properly speaking, however, nothing prevents us from supposing exactly the opposite, namely, that it was the virtuosic voice that instead required tragedy. Certainly this alternative sequence of events is easy to find throughout the subsequent course of the tragic tradition: if Hercules requires a voice, then just as surely does Nero's voice require Hercules; Callas's, Tosca. I do not mean merely that talented voices will seek out great tragic roles. Rather, I am suggesting that any strong culture of vocal performance, with or without Dionysus, has reason enough to seek—or invent—the tragic plot, complete with its motif of self-destruction, which, like the voice itself, blurs the line between agency and passivity, expression and impression. Whether this could actually be what happened in the distant Greek past cannot, of course, be known. In any case, I do not intend here to argue that what led Greek voices to tragic plots was as "banal" as what, in the wonderfully provocative suggestion of Scott Scullion, could have been enough to lead the Athenian dramatic impulse to a plot of land by the old temple of Dionysus—namely, the simple fact that this sloping spot was among the most suitable in town for large performances.[58] Nevertheless, the fact that we can plausibly contemplate a vocal rather than a Dionysian origin for tragedy should fundamentally alter our understanding of the whole tragic tradition that would follow from those lost beginnings, through to tragic opera. Indeed, I would suggest that, in essence, *there is no tragic tradition*, only the continuous (re)-discovery—aided, to be sure, by literary history—of the tragic as a vocal medium. In other words, tragedy itself no more drives this long tradition (if we can even call it a tradition) than Dionysus does. Its prime mover throughout is instead the voice.

Let us linger a while longer in Athens. The third and final element listed by Aristotle, in the *Poetics*, as essential to the tragic plot is *pathos*, a Protean term that there embraces the plot's inevitable "catastrophe" and its resulting "suffering," as well as the latter's corresponding "emotional state," simulated by the performers and shared, to a point, by their audience.[59] But *pathos* and its cognates also refer more neutrally to the fundamental "passivity" of life itself, that is, to "experience," either good or bad. We met this meaning in Chapter 1, where the same Aristotle, in *On Interpretation*, made *pathēmata* ("things felt") *en tēi psukhēi* ("in the soul") correspond to *sumbola* ("tallies") *en tēi phōnēi* ("in the voice"). There we confronted the inherent ambiguity of this last word, *phōnē*, "voice" but also "speech." Tragedy maps this ambiguity onto that of *pathos*, using experience *in extremis* to take us to and beyond the limits of language. This transgression includes the paralinguistic exclamations, prophetic babble, and the like that dot tragic scripts.[60] But it scarcely ceases once speech proper resumes. To be sure, both performers and audiences will often wind up recapturing the extralinguistic voice for meaning: groans that speak the unspeakable, tones that underscore or enervate the words with which they buzz. For a brief moment, however, the voice stands gloriously alone.

Why should this have delighted Greek audiences? The nonsemantic aspects of tragic performance surely merit some comparison to that other object of obsessive Greek spectatorship: athletics. Here too the body under pressure takes center stage. And while Olympic victors may sometimes seem to do the impossible, their strain—to say nothing of that of the vanquished—simultaneously reveals the body's throbbing limits. Like the well-trained singer, they thus offer the prospect not so much of transcending the body as of being more fully in it. Reducing singers to their words or even their melodies, therefore, is rather like reducing athletes to the rules of their various games. Eyes on the discus! (Pay no attention to the thrower's body, naked and gleaming with oil.) The truth, of course, is somewhere in between: the *discobolus* uses his body to hurl the discus, but like his sculptor Myron, he just as surely uses the discus in order to show his body, "tense as a drawn bow."[61] So, too, does the singer use his song to reveal his voice at least as much as the other way around. If performance is anything, it is a refusal to alienate the body from the work (in both senses) of art.

Back to Nero. The major Greek games included musical as well as athletic contests (the Pythians actually began as a musical competition), and on the extraordinary occasion in which he caused all four to be held in a single year in honor of his imperial visit, the emperor submitted to both kinds of trial. Variously triumphant, he was greeted upon his return to Rome, according to Dio, with the following collective shouts:

> "Hooray, Olympic winner! Hooray, Pythian winner! Augustus! Augustus! Hail to Nero Hercules! Hail to Nero Apollo! One and only tour winner! One and only ever! Augustus! Augustus! Divine voice! Happy are those who hear you!"[62]

Dio insists that he quotes this chant verbatim, and regarding the games themselves, Suetonius assures us that Nero had actually fallen from his chariot in the race in which he was declared a victor. Both reports may be true, but their repetition reveals the basic parameters of Nero's vilification during and after his lifetime. Nero is the apotheosized hero Hercules, founder of the Olympic games, and even Apollo, god of music and charioteer of the sun—except that his voice, as Dio has told us, was "small and dark," and from an ordinary rather than celestial chariot, snipes Suetonius, he fell flat on his face. The emperor, to his detractors, is ridiculous not so much for parading himself about like a god as for the invisible clothes that reveal him to be all too basely human.

Perhaps, however, there was method to this imperial madness. Chariot races, for example, were rough and tumble affairs, and Nero could hardly not have known what he was getting into.[63] Rightly or wrongly, he may have counted on impressing at least some in his audience with an emperor's willingness to get a little dirty. Hercules himself, after all, had cleaned the Augean stables, and there is evidence that Nero styled the other feat of his Greek tour, the start of construction of a canal through the Isthmus of Corinth, as a similarly Herculean labor, even moving the first earth himself—albeit, one later detractor reports, with a golden pickaxe.[64] Indeed, having styled himself as Apollo and the Sun earlier in his reign, Nero's Greek visit seems to have yielded a shift in model to antiquity's most famous strongman, who while still a mortal rid the earth of monsters, to the lasting benefit of humankind. Behind the conspicuous megalomania of such comparisons, there may well lie a deliberate program of popular appeal, analogous to that which at least one scholar has detected even in such scandalously lavish Neronian projects as the "Golden House" he constructed at the heart of Rome.[65]

What, however, of Nero the artist? It is tempting to see Hercules as a Dionysian yin to the yang of Nero's celestial Apollonianism. But Hercules is rather more than a butched-up Bacchus. In Rome, his association with the Muses was at least as old as the second-century BCE temple of Hercules Musarum, invoked by Ovid in the final lines of his *Fasti*, where Hercules plays the lyre while the Muses sing.[66] It was not the strumming Hercules, however, but the suffering one who captured the tragic imagination of Seneca and the probably different author of the *Hercules on Oeta* attributed to him. Midway through the latter play (which happens to be the longest tragedy to survive from antiquity[67]), the chorus sings expansively of another long-suffering hero, one we already have encountered: Orpheus, Greek myth's most celebrated musician. Offstage noise brings their ode to an abrupt close:

> Sed quis non modicus fragor
> aures attonitas movet?
> est est Herculeus sonus!

> But what unmeasured din is this
> that strikes our startled ears?
> It is, it is—a Herculean sound![68]

Whether we are watching a fully staged production or merely reading (or somewhere in between), this stomping entrance clearly is meant to contrast with the delicate song and dance it interrupts. Even once the thundering footsteps give way to the poisoned hero's speech, the "Herculean sound" remains markedly different from the Orphic lyricism that has preceded it.[69] That Hercules begins by calling on the Sun to reverse his chariot and plunge the world back into darkness only underscores this contrast with the arts of Apollo and his Muses. In fact, Hercules is not given a single song in the play; all of his lines are in the unaccompanied and (usually) unsung iambic trimeters that, in Greek and Latin drama alike, offered a stylized version of the natural rhythms of speech, as we already have seen.[70] The same is true in the *Hercules* of the genuine Seneca, just as it is in his model, the *Heracles* of Euripides. In other words, the "Herculean sound" is largely a distinctly unmusical one in these three tragedies, which arguably is what we would expect from the man who as a child had killed his own lyre teacher, the great Linus, after the latter beat him for being such a slow learner.[71]

However, there is a serious problem with jumping to this last conclusion, for this antilyrical Hercules is hardly alone among tragic dramatis personae. Had he been born with musical aptitude to match his brute strength, other aspects of his identity would still have tended to keep him speaking rather than singing, given the conventions of Aeschylean and Euripidean tragedy, which generally reserved song for characters who were female, foreign, or unfree, or who were either very young or very old.[72] Sophocles, alone of the three great Greek tragedians, flouts this convention, especially when his characters are "in physical pain or extreme emotional turmoil."[73] Indeed, his Heracles, in the *Trachinian Women*, does more than just sing. As Sarah Nooter puts it, "When Heracles speaks in the final episode, it is with almost every imaginable element of a poetic voice.

Deianira never sings in lyrics, but Heracles' portrayal starts and ends at an excited pitch of song."[74] Nooter's reading was partly anticipated by none other than Cicero, about to introduce, in the *Tusculan Disputations*, his own Latin translation of the song, and playing on the double meaning of *vox* ("voice," "word") explored in Chapter 2: "But let us consider Hercules himself, who is being broken by pain (*dolore frangebatur*) when he pleads for immortality through death itself: what sounds (*voces*) he gives out (*edit*) in Sophocles' *Trachinian Women*!"[75] Is this lavishly sonorous Heracles the singing exception that proves the rule, or instead a key that something *other than song* connects the suffering hero across all four of the surviving tragedies in which he stars?

The Heraclean tragedies of Sophocles and Euripides, observes Michael Silk, "may well be the first two ever on the theme of the suffering Heracles: there is in fact no clear sign of any others among the several hundred known tragedies that were produced before the end of the fifth century."[76] Contrasting this theme both with the far better attested tragic role of Heracles the "saviour" (as in, among surviving tragedies, Euripides' *Alcestis* and Sophocles' *Philoctetes*) and with the strongman's remarkable popularity in comedy, satyr-plays, and "mythological burlesques," Silk attributes Heracles' relatively rare appearances in the guise of tragic hero to his hybrid status as both human and divine, which yields "contradiction which the tragedians explore at some cost to the tragic norms and to our emotions."[77] Changing Silk's parameters to include Roman tragedy or citharoedic solo singing like that of Nero would of course alter the calculation of his rarity, but Silk's conclusion that the tragic Heracles represented, for his first audiences, an almost unthinkable hybridity and fall from grace may still help to illuminate the larger, later tradition. For as Silk reveals, Heracles' fall is not just into "all-too-human" status but very much into a human body: fleshy, fragile, and even, as Theseus is made to complain by Euripides, "womanly."[78] In *The Trachinian Women*, "humanity is crushed," while in Euripides' *Heracles*, which may be a "reaction" to its Sophoclean predecessor, "all the movement is towards humanity"; both, however, eventually present us with a broken, battered, Heraclean body (including Sophocles, who omits the apotheosis that, in the usual story, is about to save that body from the pyre's flames).[79] It is not so much that the plays reveal a demigod's latent human side as it is that they *make* a human body, disturbingly, from the body of a god: a physique conspicuous in its potential to inflict pain is traded for—or better, mingled with and contaminated by—flesh abundantly able to suffer.[80]

So dramatic is this fall into the human, mortal condition, "in a play that introduces the most scandalously embodied of Greek heroes to his capacity for suffering," that his body even becomes susceptible to the probing questions of ancient Greek medicine and ethics, as Brooke Holmes reveals in her magisterial study of "Euripides' Heracles in the Flesh."[81] Via various kinds of "symptoms," Holmes explains, which complicate the narrative presentation of the source of Heracles' madness as external and divine, the Heraclean body "erupts into visibility" in the play.[82] Let us add, however, that this body just as surely erupts into audibility.[83] That is to say, it erupts into voice, simultaneously erupting *as* voice, from within, and erupting and intruding *into* the voice, as if from outside. In this sense, the play's elaborate meditation on the insides and outsides of madness can be read as amplifications of that marvelous and mysterious aesthetic object at the very heart of its embodied performance: the actor's voice. If the human body, including the embodied mind and soul, is the medium of tragic suffering, then the play's fall-into-body promises its audience not just that medium's inscription, but (seemingly) its very manufacture. This, at least, is what a talented actor must have tried to capture. To be sure, he was spared the task of "bellowing like a bull"[84]: Heracles' madness unfolds offstage, and its visual and audible symptoms are described to us by Lyssa, who personifies insanity itself. In part, the play thus avoids what would have been difficult and, more to the point, grotesque to stage, since its direct representation would have risked crossing into the animal-sound habitat of comedy—where the tragedians themselves, in Aristophanes' *Frogs*, can be made to look and sound like bellowing tragic madmen.[85] But Lyssa's indirect representation of sound also primes our ears for what is to come, readying us to listen through and beyond language, even as Heracles returns to the script. We again hear Heracles when he again begins to hear himself, and what we hear is not a speech but a voice, one in which suffering and self-awareness are, achingly, the same thing.

Clearly, the performative pressures on any actor playing Heracles would have been considerable, also because, by the conventions of the Athenian stage, he already would have played, earlier in the same play, the part of Lycus, the usurper king of Thebes who orders the slaughter of Heracles' family, a deed which Heracles himself, ironically, will carry out. The actor's primary vocal task was not, however, impersonation in some narrow sense, giving each character a distinct vocal identity. Generally speaking, that kind of vocal realism can hardly have been the point in tragedy, given, *inter alia*, the number of plays with female protagonists, all played by men. Furthermore, the *Heracles* provides an excellent example of how the playing of two parts by the same actor can help to blur, deliberately, the difference between the respective characters:

> The fact that the same actor played both Lykos and Herakles would seem to emphasize, in terms of simple stage-craft, the bond which joins the two figures despite their apparent dissimilarity. We can surmise that the audience must have been able to recognize who doubled what roles in a production since prizes were awarded to actors in the competitions at the drama festivals; the extraordinary prestige of the protagonists, who alone competed, as well as the distinctive virtuosity of each of them in gesture and vocal delivery, would, in any case, have served to focus the spectator's attention upon the entirety of an actor's activity in the presentation, just as a modern audience will follow a singer through his various roles in an opera.[86]

If the same actor must give voice to two characters, Lycus and Heracles, then just as surely must his voice reveal the latter's personality to be a divided one, the murderous extreme of which has already been mapped by its anticipation in the former. In more than one sense, the actor's voice must work here to conjure, without fully separating, multiple personalities. It all must have been quite extraordinary to hear. In precisely this regard, however, the text we read is anything but mute, not least because, as Christina Kraus has observed, the play is bursting with typically Euripidean wordplay, especially puns and paronomasia, which again and again cause the characters to say one thing while sounding almost like they are saying another.[87] For the actor these are irresistible opportunities for vocal virtuosity, but here too there is no discernable difference between script and record: even the silent reader, much like Heracles himself, cannot help hearing voices in this play.

Earlier we saw that the Senecan and pseudo-Senecan interest in Hercules as a tragic figure is driven by the unspeakability of his suffering, whether psychic, as he slowly realizes that he has murdered his family, or somatic, as the centaur's poison consumes his flesh. Their Euripidean counterpart has brought home a complementary lesson: even in an aesthetic context, like that of Greek tragedy, in which music is key, the failure of speech does not always or immediately provoke a musical solution. To be sure, the Chorus often picks up the slack in such cases, as it does in Euripides,[88] and as it will do in *Hercules on Oeta*, where it responds to the hero's lament with a song (or at least, what purports to be a song) that begins, "What could pain (*dolor*) not conquer?"[89] The unspeakable pain itself, however, we already have heard in the very voice of the play's protagonist.

Indeed, that voice must succeed in making us hear that pain regardless of whether it speaks (as in Euripides, Seneca, and pseudo-Seneca) or sings (as in Sophocles). One might be inclined to say that each of these plays requires the voice in order to "make sense"—except, of course, that any such sense again and again seems to lie in the very senselessness of suffering. In the words of pious incomprehension with which the Sophoclean chorus, addressing its leader, looks back over and closes *The Trachinian Women*:

> Maiden, do not be left in this house.
> You have seen strange, terrible deaths,

> many disasters, till now unknown—
> and none of these things is not Zeus.[90]

In the end, the play's suffering has not meant anything at all, or at least, not anything we are meant finally to understand. For Athenian audiences, this fact represented neither a dramatic failure nor a sacred mystery. They did not go to the theater in search of knowledge, any more than they expected to "see" terrible deaths, which normally happened offstage. They instead went, foremost, in order to listen. Far from being an impediment to their listening pleasure, ineffability was one of the things that helped them to hear the actors' voices, as voices. To be clear, I do not so much mean that the voice expresses the ineffable as the opposite: namely, that ineffability is a medium for the expression of the voice itself, as tragedy's central object of attention.

Let us extract from tragedy's suffering sounds a principle, which we shall call "the Heraclean." By this we shall designate the voice that, even when speaking or singing, is expressive in a way that is neither linguistic nor musical (at least, not in the ordinary modern meaning of these terms).[91] Seen analytically, the Heraclean is thus the expressive remainder after language and music have been accounted for. In what way is this remainder expressive? The Heraclean is not semantic in the way of language, but neither does it offer music's prospect of the aesthetic unfettered by meaning. In other words, to the extent to which we understand them both in structural terms, language and music bracket and thus help to locate the Heraclean but do not themselves embody it. On the question of meaning, the Heraclean offers instead a kind of deferral, exposing without explaining the very flesh (ensouled and envoiced, as we saw in Chapter 1) in and from which it emerges. The Heraclean is the body under pressure, the body as resistance, the body itself as medium, expressing as it is impressed. The Heraclean, we may even say, *is* the body, as it becomes audible in tragic speech and song. In this regard, we could extend its purview beyond the voice to include the body's visibility in tragic gesture and dance, no less a part of tragic *mousikē* than word and music.[92] In all these ways, the Heraclean body looms large in tragedy—and not only when wearing the mask of the hulking hero himself.

Playing on the ambiguous relationship between experience and agency already implicit in his band's name (The Jimi Hendrix Experience is as much our experience as it is his), Hendrix rephrases his title question as, "Have you ever been experienced?"[93] He then answers the revised version as a question posed to himself, less with what he says—"Well, I have"—than with the soulful guitar solo that he introduces with "Let me prove it to you." A year before the album's release, Hendrix is reported to have remarked, "I want to do with my guitar what Little Richard," for whom he had played, "does with his voice."[94] Indeed, in Hendrix's hands the electric guitar becomes comparable less to the ancient lyre than to tragedy's *aulos*, the reed-based wind instrument that, according to Aristotle, so resembled the voice as to provide far better accompaniment to solo singing than the lyre and that, according to Pindar, Athena herself had invented to imitate the mourning voices of the sisters of Medusa.[95] (Compare the wailing sounds of the modern clarinet, a far closer analogue than the flute, though the latter is a common mistranslation of *aulos*.) The point, once again, is not that the sound of experience is necessarily musical but, rather, that it is so potentially independent of language as to be expressible (or at least to seem to be so) by a "voice" that cannot speak. Hendrix's guitar and Athena's *aulos*, which she tossed away in disgust after seeing how playing it made her cheeks bulge, do not offer disembodied abstractions of speech but, rather, prosthetic extensions of the voice's deep tissue. (Indeed, in objecting to *aulos*-training for young aristocrats, Aristotle singles out the way in which the instrument itself diverts the mouth from the articulation of words.)[96] It is deep down in that vocal flesh—the Aristotelian heart of the matter, as we saw in the introduction—that the roots of the Heraclean are to be sought: not merely the experienced voice but the voice of experience itself, the voice that feels, the voice that has felt.

Indeed, we are reminded of the first chapter's "Aristotelian" Sirens, who gesture at their own resonating flesh and who, in the funerary monument to which they once belonged, mimed the work of the dead poet's mourners. "Every tragedy deals with the staging of mourning," observes Nicole Loraux in her extended essay, *The Mourning Voice*, which offers a proposition we are pleased to repeat: "I believe we have no choice except to return to the question of voice in tragedy."[97] To locate the voice in question, Loraux turns to the "connection, at once conflictual and constitutive, between *logos* and *phōnē*. Conflict, let me emphasize, rather than coexistence."[98] Perception of this conflict, as Loraux notes, is at least as old as Plato, who famously banishes the *aulos* and other wind instruments from his ideal republic, along with "laments and mournful wailing," preferring instead the stringed "instruments of Apollo" and music that accompanies without rivaling speech.[99] Loraux instead follows Nietzsche, who in *The Birth of Tragedy* observes the same conflict but prefers—along with tragedy itself—"Dionysian music" (which really is "music generally") to the "architectonics in sounds" of Apollo's strings.[100] Under the same Nietzschean rubric of Dionysian music, Loraux subsumes the inarticulate cries of the tragic mourner: "In almost every case, even if the cry dominates, music, whether it be soft or loud, is evoked at the same time..."[101] And so, "in the tragic world all moaning tends to consider itself music."[102] The conflict between *phōnē* and *logos* thus turns out to be Nietzsche's, that is, one between music and speech, and "the question of voice in tragedy," accordingly, is best asked of the former: "By inviting the reader to listen to the mourning voice in tragedy I wish not only to emphasize listening over seeing in theatrical representation, but also song over discourse (*logos*) and to heed the lyric passages over the iambic meter of dialogue, in other words to focus more on the chorus than on the protagonists."[103] Nietzsche before her had hoped to effect a similar shift of attention: "The genesis of Greek tragedy now tells us with great clarity and definiteness how the tragic work of art of the Greeks was truly born from the spirit of music; we believe that, with this thought, we have done justice for the first time to the original and quite astonishing significance of the chorus."[104]

Taken to an extreme, such a correction banishes the great Bernhardt and her ancient counterparts from the tragic stage. Indeed, this is precisely what Nietzsche does, rejecting everything from Euripides, whose speaking characters usurp the centrality of the Aeschylean chorus, to operatic recitative. But Loraux's somewhat forced reading of tragic wailing as music unwittingly reveals where she and Nietzsche both go wrong. It is not music that is the soul of tragedy but, rather, the voice, whose Heraclean spirit inhabits speech and song alike. To be sure, music places special pressure on the vocalizing body. But so too does spoken ambiguity, in all of William Empson's "seven types": when *nescioquid* and the like give us hermeneutic pause, we confront a particularly poignant kind of vocality. (That said, the voice is also there even when we say, plainly and simply, precisely what we mean.) Even tragedy's ever-present *aiai*, "the cry for all cries" for Loraux, hardly "introduces us to a world in which there is no meaning other than sound itself," since anyone who has suffered knows all too well where such sounds come from.[105] In any case, as we already have suggested, what brought the Greeks to the theater was neither an escape from meaning nor a full surrender to it. Tragedy does not offer us another world at all. Rather, it stages the world we already know, the world of experience itself, inhabited by bodies and souls—and thus also voices—that are expressive of experience in nearly all the ways a human being can be, at least from a distance.

What, then, of tragedy as literature, that is, as a text? *The Birth of Tragedy* does not really tell us what we are to do with surviving works as musical artifacts, but the matter was very much on Nietzsche's mind, for he devoted his few remaining years as a professional classicist, before resigning his position at Basel, to the study of ancient rhythm and meter.[106] Certainly what we saw in the last chapter regarding Greek lyric is equally applicable to the sung parts of tragedy: there is more ancient music notated here than first meets the modern eye. Any sense of deficiency diminishes further if we regard these scripts not only as musical artifacts, but more generally as vocal ones.

Such considerations, however, may miss the fundamental point of the tragic phonograph. What matters about a tragic text is not so much the groove as the vinyl, that is, not what exactly it records, but the very fact of its impressibility. The medium is not only the message here. It is the Tragic Muse herself. Text into body into text—so goes the early history of every surviving tragedy, from script to performance to circulation as literature (a sequence that makes tragedy exemplary of rather than exceptional to ancient literature generally). Situated between two acts of writing, the actor's body is no extratextual intermission but is itself a medium, a body whose innate and acquired modes of expression have been primed to render as audible as possible the extraordinary pressures of extraordinary circumstances. This is what the Greeks went to the theater to hear, and there is no reason to suppose that their motives were any different as readers. A text did not really give them the voice of Hercules, but neither did an actor. Both offered instead, in one form or another, something smaller but infinitely more compelling: the sound of experience itself.

So, one last time: "Are you experienced?" At the end of this long trip through the tragic tradition, we know why its voice wound up here, with a question answered not only by an instrumental solo, but by one that was actually recorded backward in order to produce its distinctive feel: a reply, in other words, that has never been language and, even more important, has never been anything *but* a record. The experienced ear sought much the same when the music came instead from speakers pointed out from the stage and orchestra of the Theater of Dionysus, transmitting, through the conical amplifiers of the tragic mask,[107] the sounds of suffering long past. And what the voice of experience, whatever its phonograph, has forever been telling us is this: singly, we may be defined by our words and deeds, or even by our high or low notes, but what we share is our fundamental passivity before a world that is both horribly and magnificently beyond our control. Even Nero knew that. When it comes to what life feels like, we are all experienced—not necessarily mad, and, as Hendrix concludes, "not necessarily stoned, but beautiful."

TRACK FIVE

Amazing Grace

Amazing grace! (how sweet the sound)
That sav'd a wretch like me!

—JOHN NEWTON, "AMAZING GRACE," 1779

"What is this sound—so loud, so sweet—that fills my ears?"

—CICERO, "SCIPIO'S DREAM," *ON THE REPUBLIC* 6.18

We shall approach antiquity's most famous voice from a distance, beginning some fourteen centuries after the brutal dismemberment of the body to which that voice had once been attached. We are in Verona with Petrarch, in the summer of his fortieth year, surrounded by medieval towers perched atop the still hulking remains of the city's Roman past. We find him in the library (as we did Julius Pollux), among the dusty books of the cathedral's chapter house. Here the scholar and poet uncovered at last the text that had eluded his meticulous searches throughout Europe: Marcus Tullius Cicero's collected letters to his friend and confidant, Titus Pomponius Atticus.[1] The discovery would prove to be a watershed in literary history, not least because no other ancient text would exert as pervasive an influence on the Latin of the emerging movement we call the Renaissance.

When Petrarch had finished reading, he drafted a letter of complaint, addressed to none other than Cicero himself. Sometimes misread as genuine condemnation, the letter is instead full of tender irony, as the man who has been called the father of humanism pretends to reproach his most beloved author precisely for having been human. In the letter's opening lines, this story of books and libraries yields to the language of living sound:

> Francesco to his dear Cicero: Greetings. With enormous eagerness I have read through your letters, which I sought far and wide and found where I least expected to. I have heard (*audivi*) you saying (*dicentem*) many things, bemoaning (*deplorantem*) many things, changing many things, Marcus Tullius, and having long known the kind of teacher you were for others, I now realize who you were for yourself. In return, hear (*audi*), wherever you are, what I can no longer call advice but, rather, a lament (*lamentum*), born of true affection, which one of those who came after you, devoted to your glory, pours out not without tears.[2]

We often imagine that our texts "speak," and Petrarch's language here underscores the vividness with which Cicero's intimate letters have just made their author present to their distant reader (and vice versa, given the conceit of a reply). Nevertheless, such noisiness would perhaps be more natural in response to Cicero's orations or dialogues, which pointedly invite us to hear him "saying many things"; a letter, by contrast, as its very name suggests, does not ordinarily pretend to be anything other than a written text.[3] So, let us ask what exactly Petrarch thought he "heard" in Cicero's letters, there in Verona in the summer of 1345.

For a first answer, we must briefly detour south, and back to 58 BCE, when, in March and April, Cicero traced an erratic path down the Italian peninsula. Along the way he dispatched a series of seven remarkable letters in the opposite direction, toward Atticus in Rome, though he was not entirely sure his friend was still there and entertained vain hopes of being joined by him on the road. Except for the last of these letters, written from the port of Brundisium, where Cicero would board a boat for Greece and exile, each runs no more than a paragraph. This brevity, however, hardly prevents them from overflowing with their author's anguish. In the fourth or fifth letter (their precise order is not always clear), Cicero bemoans his wretched life and its torments (*ego vivo miserrimus et maximo dolore conficior*) but then, at the end, briefly rallies:

> I ask you only this: Since you have always loved me for who I am, don't let that change. For *I* have not changed. My enemies have deprived me of my possessions but not of who I am.[4]

This letter, along with more than four hundred others to Atticus, eventually made its way into the posthumously published corpus, which Petrarch not only read but also transcribed, from start to finish, doubtless in the careful hand for which he is known, though his copy, like his model, was itself long ago lost. He therefore could hardly have missed a feature of the lines just quoted that my translation from Latin has obscured: Cicero's utter surrender to alliteration and other consonance. This begins with the letter *t*:

> **T**an**t**um **t**e oro u**t**

But it soon switches to *m*:

> quonia**m** **m**e ipsu**m** se**m**per a**m**asti, ut eode**m** a**m**ore sis; ego eni**m** ide**m** su**m**. Ini**m**ici **m**ei **m**ea **m**ihi, non **m**e ipsu**m** ade**m**erunt.

The particular letters of the alphabet in play are scarcely an accident, since the start of the passage is addressed to you (*te*), but the remainder is all about me, me, me (*me*, *ego*, *mei*, *mea*, *mihi*, *me*), a balance that an uncharitable critic might well call emblematic of the entire collection, in which Cicero seems so often to be talking mostly to himself that one hardly misses Atticus's lost reply.[5]

This, however, is Cicero—and Ciceronian sound—at their most cartoonish. Far earlier, Petrarch had been captivated by more dignified strains, as he would later reminisce:

> Siquidem ab ipsa pueritia, quando ceteri omnes aut Prospero inhiant, aut Esopo, ego libris Ciceronis incubui.... Et illa quidem etate nichil intellegere poteram; sola me verborum dulcedo quedam et sonoritas detinebat, ut quicquid aliud vel legerem vel audirem raucum mihi longeque dissonum videretur.

> Beginning, in fact, in my very childhood, when everyone else gapes at Prosper or Aesop, I snuggled up to the books of Cicero. . . . Of course, at that age I couldn't understand anything; what got my attention was simply a kind of sweetness and sonority of his words. The result was that anything else I read or heard sounded very different, and harsh.[6]

Telling are the terms in which Petrarch describes the arresting beauty of Cicero's words: *dulcedo* and *sonoritas*, "sweetness" and "sonority," perhaps best understood in hendiadys as meaning "sweet sound," along the lines of Shakespeare's "sound and fury." It is the synesthetic first term that is especially revealing. For although we can easily imagine any reader enjoying grand Ciceronian moments of coincidence between sense and *sonoritas—O tempora! O mores!*—rarer are the passages where we can truly say that Cicero's sense is itself "sweet." *Dulcedo* refers instead to sound at least partially independent of sense, which is precisely what Petrarch has already suggested by telling us that he felt this precocious attraction well before he was old enough to "understand" what he was reading.[7]

This boy who turned his ear to Cicero no sooner had grown up than he found himself looking back on Laura, the woman he had briefly loved and forever lost. Petrarch opens his collection of sonnets about her with this remarkable address to the reader:

> Voi ch'ascoltate in rime sparse il **suono**
> di quei sospiri ond'io nudriva 'l core
> in sul mio primo giovenile errore
> quand'era in parte altr'uom da quel ch'i' **sono**:
>
> del vario stile in ch'io piango e ragiono,
> fra le vane speranze e 'l van dolore,
> ove sia chi per prova intenda amore,
> spero trovar pietà, non che perdono.
>
> Ma ben veggio or sì come al popol tutto
> favola fui gran tempo, onde sovente
> di **me medesmo meco mi** vergogno;

e del mio vaneggiar vergogna è 'l frutto,
e 'l pentersi, e 'l conoscer chiaramente
che quanto piace al mondo è breve **sogno**.

> You who listen to the sound, in scattered rhymes, of those sighs on which I nourished my heart, in the waywardness of my first youth, when I partly was a different man from the one I now am: where there is someone who knows firsthand what love is, I hope to find sympathy, as well as forgiveness, for the changing style in which I weep and speak, wavering between vain hope and vain suffering. But I see full well now how I have long been the talk of everyone, and this makes me ashamed of myself, by myself, over and over. And shame is the fruit of my vanity, and remorse, and clear knowledge that all that pleases the world is a brief dream.[8]

Petrarch's Latin title for the collection promises us "Fragments of Things in the Vernacular" (*Rerum vulgarium fragmenta*); that these are foremost pieces of sound is clear from their reinvocation, in the opening poem's first line, as *rime sparse* ("scattered rhymes"). These *rime* are, of course, by metonymy, the single poems gathered here into this corpus of love and loss, but this same tension between collection and dissipation is realized, in microcosm, in the literal "rhymes" of the opening poem. Here the elegant symmetry of the Petrarchan sonnet is both amplified and disrupted by the presence of that most basic (and disorienting) kind of rhyme, indeed, the very kind of rhyme which the *membra disiecta* of Orpheus, plainly evoked here, produced as they floated down the Hebrus River: an echo. For the "Eurydice . . . Eurydice . . . Eurydice" of Virgil's Orpheus, and the repeated *flebile nescioquid* of Ovid's, Petrarch substitutes *suono . . . sono . . . sogno*, "sound . . . I am . . . dream": a trilogy that takes us outside of what the poem strictly says and into its oneiric blending of music and me. To this sonorous self-consciousness belongs a second echo, deep inside the poem: *me medesmo meco mi vergogno*. Here, the aggressively Latinate *meco* seems to mimic Petrarch's normal (i.e., Latin) language of self-reflection, in which he is quite fond of *mecum*, and the whole formulation, of course, sounds suspiciously like something we just heard in Cicero. What Cicero had used to rally his pride, Petrarch turns to the shame (*vergogno/vergogna*, another echo) we are invited to hear in his "sighs" (*sospiri*) and "weeping" (*piango*)—sounds that, however, betray a host of other, deeper, and more ambivalent emotions. The latter, *piango*, "I weep," is contrasted to *ragiono*, a pointedly intellectual term for "I speak"; together they comprise the double work of Petrarch's *stile*, "pen" or "style." Our readerly work, he is telling us, should be double too, attentive both to sound and to sense. The poet's first line, however, echoed by his last, is unambiguous about which to prefer: "You who listen, in scattered rhymes, to the sound . . ." Little wonder that these "fragments" are better known by the more insistently musical title given them by later readers and imitators: the *Canzoniere*, or "Songbook."

Petrarch never suspected that his sonnets would guarantee his place in literary history. He was far more concerned about the reception of his Latin poetry, especially his epic *Africa*, over which he labored sporadically throughout his adult life. The nine-book poem tells the story of the conquest of Carthage in the Second Punic War by the Roman general Publius Cornelius Scipio, called thenceforth Africanus. In the third book, Scipio sends his friend and lieutenant Gaius Laelius to the bejeweled palace of the Numidian king Syphax to propose an alliance. Laelius is warmly received and, after dinner, is treated to a song:

> Vixdum finis erat dapibus, quum comptus in ostro
> astitit ante oculos iuvenis, patrioque canoram
> increpuit de more lyram; dulcedine mira
> obstupuere omnes; sonitum mox verba sequuntur.

> Just done with the feast, we looked up and saw,
> a purple-clad boy who struck up a tune
> in an ancestral mode on his lyre—so sweet
> that all were amazed; then words follow sound.[9]

Like many such figures in epic from Homer onward, this musician is a stand-in for the poet himself, the real source of the verses about to be sung. In introducing his alter ego, Petrarch offers some fleeting but remarkable glimpses into the nature and limits of his own craft. The banquet would have been a full-on sensory experience, but Petrarch (rather like Gellius in Chapter 3) leaves our plates and wine cups empty, whetting instead our appetite for words by telling us that the seating arrangements placed "honey-tongued" Laelius next to his royal host. A line later, however, the banquet is over, as the conversation we have only barely had time to begin to imagine is interrupted by the jarring appearance, described in aggressively visual terms (*astitit ante oculos*), of a yet unexplained boy dressed in purple. This moment of startled silence quickly gives way to equally startling (*obstupuere*) music of which, frustratingly, we ourselves cannot enjoy the "sweetness" (*dulcedo*, which of course we have seen before). Not to worry, Petrarch reassures us, "soon words follow the sound," which plays on two meanings of "follow": the words come next (both in the concert and in the poem), but they also keep pace with the lyre's tune. What tune was that? Here Petrarch offers a kind of riddle. The boy plays the lyre *de patrio more*, "in the fashion of his fathers," which notionally would mean that his music was Numidian. But Petrarch, the real bard here, is also following an ancestral mode, offering the words to come (like those of the entire poem) in the dactylic hexameter of the ancient Roman (and Greek) poets of whom he has made himself heir. We cannot hear the boy's lyre or its African strains, but this other lost music is there to be heard in Petrarch's every line.

That music, as every beginning student of Homer or Virgil must be taught, was made not of stress-based rhythm but of the long and short syllables of quantitative meter. But by Petrarch's day, spoken Latin had largely lost its classical vowel lengths, and written copies even of Virgil followed the rules of medieval orthography, which had long ago substituted, for example, a simple *e* for the long diphthongs *ae* and *oe*; nor were there any dictionaries to help sort things out. Thus, while classicizing Latin poetry in Petrarch's wake would progress so rapidly that soon he himself could be criticized for his errors of prosody, only a few poets in the centuries before him had come anywhere near his level of success. This, indeed, was a large part of the justification for the elaborate publicity stunt by which, on Easter Sunday, 1341, before a distinguished audience assembled on the summit of Rome's Capitoline hill, Petrarch was crowned the first (it was claimed) "poet laureate" since antiquity.

For the occasion, Petrarch delivered a finely wrought acceptance speech modeled on Cicero's oration *Pro Archia*, which he himself had tracked down eight years earlier, in Liège. He tells the (literally) colorful story of its rediscovery in the very same letter in which he describes his boyhood attraction to Cicero's *sonoritas* and *dulcedo*:

> Whenever I was off on long sight-seeing trips, something I did often back then, and I spied ancient monasteries from afar, I made an unplanned stop then and there. "How do I know," I would say, "whether one of the things I am looking for is here?" When I was about twenty-five, I was hurrying through Belgium and Switzerland and had come to Liège, and having heard that there was a nice stash of books there, I stopped and kept my traveling companions waiting until I had copied two orations by Cicero, one by a friend's hand and the other by my own. You'll laugh, but in that excellent barbarian polity it was serious work to find any ink, and what I did find was yellow like saffron.[10]

Having undertaken to defend the poet Archias, whose Roman citizenship had been questioned by a politically motivated prosecution, Cicero had used the opportunity to mount a lively defense, with an Orphic ring, of the poetic profession itself: "Deserted cliffs and crags give an answering echo to the voice; wild animals are often stopped in their tracks by the sound of singing; should we, educated as we are, fail to be moved by the voice of poets?"[11] Cicero even goes so far as to credit the formation of his own voice, "if ever it has been of service to anyone"[12]—a heavy-handed wink above all at the momentous oratory of his consulship the year before—to the early encouragement and advice of Archias himself, though he keeps a discreet silence about the poetic ambitions of his own youth, soon to be reprised in the ill-fated *On His Consulship*.

Though often overlooked, part of what binds Petrarch so tightly to Cicero is the fact that Cicero too was primarily a prose author who also cultivated a poetic voice, however unkind posterity would be to the latter (a fate which Petrarch would himself share, at least regarding his Latin poetry). This is why it made perfectly good sense for Petrarch to accept the laurel crown he won as a poet with a speech in Ciceronian prose. Let us briefly consider a particular consequence of these sympathies between Petrarch and Cicero and between prose and poetry. Petrarch's relative mastery of the syllable lengths of ancient poetry primed him to hear, not in Cicero's verses but in his prose, something that had largely been lost on centuries of medieval readers who had come before him: namely, his extensive use of prose rhythm.

Cicero was one of antiquity's most celebrated practitioners of the originally Greek habit of marking, with specific patterns of long and short syllables, the ends not only of complete sentences but also of the syntactical and rhetorical units of which they were composed. In the passage from the *Pro Archia* from which I have been quoting, syllables belonging to such patterns (generally called *clausulae*) appear in boldface:

> Saxa et so**litudines** vo**ci respondent**, bestiae **saepe immanes** can**tu flectuntur atque consistunt**; nos **instituti rebus optimis non poetarum voce moveamur**? Homerum Colophonii civem **esse dicunt suum**, Chii **suum vindicant**, Sala**minii repetunt**, Smyrnaei vero suum **esse confirmant**, itque etiam delubrum eius in oppido **dedicaverunt**, permulti alii praeterea pugnant inter se **atque contendunt**.[13]

As John Dugan notes, "For Cicero, prose rhythm knits together prose to give a bodily integrity and force that are the antitheses of the slackness and bloatedness imputed to his prose by his Atticist critics; and it also allows the text to have a breath of life like that of a living performance. Cicero thus presents rhythmical prose as crucial to the long-term survival of an oratorical text by replicating the grain of the orator's voice in performance."[14] But in Late Antiquity and the Middle Ages, as quantitative pronunciation faded and natural vowel quantities were lost, the complex quantitative patterns of classical prose went the way of quantitative verse. They were replaced by patterns based instead on stress accent, eventually organized into a streamlined system called *cursus*, which would become a key component of the High Medieval rhetorical education received by Petrarch. Medieval readers trained in *cursus* knew that ancient prose sounded different and even knew why, but few before Petrarch were in a position to replay that lost soundtrack with any accuracy.

Petrarch managed to pull somewhat away from *cursus* in his most classicizing prose,[15] but he was not in a position to imitate the sounds of his ancient models, in part because the ancient system was so complex (its full reconstruction, centuries later, is regarded as a monument of German classical scholarship), and in part because he lacked unmutilated copies of Quintilian and of Cicero's *Orator*, the two major Latin treatments of the subject. Petrarch's position is thus vaguely analogous to that of ancient audiences in the Roman Forum who, we are told, were disposed to cheer a well-turned *clausula* though not necessarily able to produce the same themselves. In a different light, we may perhaps compare the impact that fourteenth-century music had on the formation of the Petrarchan sonnet, though Petrarch himself was no composer.[16] In various ways, in other words, what Petrarch wrote reflected without directly reproducing what he could hear.

Can we go one step farther and trace a long stylistic arc from the Ciceronian *clausula* to the musicality of the *Canzoniere*? Such a suggestion at first seems preposterously reductive, flattening out differences of language, genre, period, content. Still, the history of literary style has left us with a paradox, connecting Cicero and Petrarch precisely in terms of these seeming antipodes. For Cicero would continue to be the Renaissance's leading model for Latin prose, but when time came to enthrone a similarly "Ciceronian"—that is, single and authoritative—model for Italian poetry, the clear choice was

Petrarch.[17] The virtues celebrated in each would be long to list, but I would like to suggest that their selection as models is due not only or even primarily to the aggregation of their signature elements, such as Cicero's favorite *clausulae* or Petrarch's preferred rhyme schemes. Rather, Cicero and Petrarch offer a battery of stylistic options that add up to something like a high-volume mark for *sonoritas* itself. You can write Latin prose or Italian verse that sounds *different* from Cicero's or Petrarch's, but it is not easy, within the basic limits of decorum and sense, to make either medium make *more* sound than these two do. In other words, it is not so much the distinctiveness of their respective voices that sets them apart from others as, rather, the omnipresence of sound in their works.

What would it mean to come at Renaissance literature as, foremost, a vocal problem, that is, as a reevaluation of literary language as an acoustical medium? Of course, one could object that talking about literary voice is just another way of talking about literary style, and that there is hardly anything new in understanding the Renaissance in stylistic terms. But if voice and style can sometimes boil down to the same things, the way we understand their stakes can be quite different, as we already have had occasion to consider in this book. Above all, one acquires a style only long after learning to write (if at all), but in some basic sense, our voices are ours even before we learn to speak. However much a literary voice may be the product of stylistic polish, its notional roots are deep in our tissue—so deep, in fact, that Cicero's "I am who I am" in the letter to Atticus we were reading a moment ago can sound almost like the infantile babble we considered in Chapter 2. Deeper still, for Petrarch, this same babble strikes nothing less than the (vocal) chord of our shared humanity.

If Renaissance Ciceronianism began in the rediscovery of ancient ways of mediating the human voice, it would end in the parade of later humanists ridiculed by Angelo Poliziano as being not humans at all, but mere "apes of Cicero," targets too of Erasmus's blistering satire, the *Ciceronianus*, which repeats (as does Poliziano) and expands Quintilian's far earlier derision of those who think themselves "Cicero's natural-born brothers," just because they end a sentence with *esse videatur.*[18] But the sonic dimensions of the recurring Ciceronian obsession teach us something about Cicero's own phonographic project in the first place, enabling us to retrace, in a new way, his influence already in antiquity, and beyond. That Cicero succeeded in representing *a* voice—received as his own—is clear from what his imitators, both serious and silly, understood themselves to be doing. However, to consider this achievement only in stylistic terms is to miss an accomplishment we might more properly call technological: namely, the capturing of *the* voice, plain and simple. Indeed, in Cicero's case, the former cannot really be understood without the latter. His signature voice may be more than the sum of its parts, but it nevertheless requires that sum, namely, the voluminous aggregation of what he found he could represent of and as the human voice, in a written text. This is the sound that kept him on the air, here and there, even after medieval schools stopped manufacturing ancient media players—and which in Verona again came though, loud and clear, to thrill and astonish Petrarch. But long before any of this could happen, Cicero himself had to lose his voice, as a closer listen will now reveal.

Once upon a time, Cicero had made his own efforts to tune in the rhetorical past. He was, first of all, an avid reader of Greek and Roman oratory; regarding the latter, he could claim to have "tracked down and read" no fewer than 150 speeches by Cato the Censor alone.[19] But it was instead as a writer that Cicero found his most important relationship to bygone voices. In the mid-50s, back from exile and resigned, for the moment, to political retirement, he penned his first major treatise, *On the Orator*, framed as a conversation principally between Lucius Licinius Crassus and Marcus Antonius, set in the former's villa in the hills outside Rome, in September 91, a few days before the death of Crassus and a few years before that of Antonius. Cicero would have been sixteen at the time, and though he knew both men (Crassus was an early mentor), he does not pretend to record a dialogue he himself had heard:

> Nos enim, qui ipsi sermoni non interfuissemus et quibus C. Cotta tantummodo locos ac sententias huius disputationis tradidisset, quo in genere orationis utrumque oratorem cognoveramus id ipsum sumus in eorum sermone adumbrare conati; quodsi quis erit, qui ductus opinione volgi aut Antonium ieiuniorem aut Crassum pleniorem fuisse putet, quam quo modo a nobis uterque inductus est, is erit ex eis, qui aut illos non audierint aut iudicare non possint; nam fuit uterque, ut exposui antea, cum studio atque ingenio et doctrina praestans omnibus, tum in suo genere perfectus, ut neque in Antonio deesset hic ornatus orationis neque in Crasso redundaret.
>
> I myself was not present for the conversation; rather, Gaius Cotta related to me its outline and highlights. I therefore have tried, in their conversation, to conjure each orator in the precise style of speaking in which I had come to know him. But any readers who, misled by common opinion, believe either that Antonius was leaner or Crassus fuller than I have presented them here must be counted among those who either did not hear them or cannot judge. For as I explained before, each outshone his peers in effort, talent, and learning and, at the same time, achieved perfection in his own particular style, with the result that our subject, rhetorical embellishment, was neither lacking in Antonius nor excessive in Crassus.[20]

Cicero's claim to have reproduced the distinctive style of each orator is largely belied by the three books of dialogue that follow. His real point, A. E. Douglas suggests, is rather different:

> Cicero is not here making a claim to historical accuracy so much as, first, explaining to the literal-minded why both characters sound so much like Cicero himself, and, secondly, seizing the chance to dilate on the excellence of his two predecessors, about whom, incidentally, no other ancient critics seem to have been nearly so enthusiastic. They were not as good as Cicero, but Cicero will let the will pass for the deed and make them sound like him.[21]

It is not that there is no difference at all between the dialogue's two principal voices: although both may sound Ciceronian, "lean"

Antonius is somewhat measurably less inclined to use a favored figure or rhythm than "full" Crassus. The distance between them, however, can be mapped onto Cicero's own range within and between the speeches of his published oratorical corpus.[22] In other words, rather than Crassus and Antonius in high fidelity, the dialogue gives us pure Cicero—in stereo. By the end, Cicero's decision to give the starring roles of *On the Orator* to his two famous predecessors has merely set the stage for an almost Oedipal muting of them both.

But while the historical voice of Crassus would surrender utterly to that of his protégé (for we know it from no other source), that of Antonius would not be so readily silenced. Indeed, in this book, we already have heard it, or at least have tried to, as we struggled also to hear the identically described "dark voice" of Rome's infamous singing emperor. It is Quintilian, writing nearly two centuries after Antonius's death, who tells us that the orator's voice was *fusca*. Quintilian's source, which had instead called it *subrauca*, "fairly hoarse" or "harsh" or "husky," was none other than Cicero himself, though not in *On the Orator*, where, remarkably, the matter is never mentioned, even in the gushing résumé Crassus provides of his interlocutor's signature style:

> Do you see the style that belongs to Antonius? Strong, impassioned, energetic in its delivery, forearmed and buttressed on every side of his case, keen, penetrating, unencumbered, giving each matter its due, yielding honorably, pursuing ferociously, alarming, imploring, with ever changing eloquence, till our ears cannot get enough (*nulla nostrarum aurium satietate*).[23]

Our ears get the missing detail only in the *Brutus*, written a decade later and framed as a dialogue not of the dead, but of Cicero himself with his title character and dedicatee, in the company of Atticus. The famous orators of the Greek and Roman past are the subject of this more recent conversation, which turns eventually to Antonius and Crassus. Cicero starts with the former, who, he first tells us, arranged his thoughts as a general arranges his army; his prodigious memory allowed him to prepare carefully for speaking, yet he always appeared to speak *ex tempore*. His Latin was not fancy but was the sort that befits an orator, something Cicero revealingly insists is not really an oratorical virtue at all but, rather, should be proper to every Roman citizen ("it less glorious to have good Latin than it is disgraceful not to," he quips), providing a foundation for higher talents. These follow in a complicated pair of comparisons to Demosthenes that conclude the portrait and that bear quoting in full:

> Nevertheless, in choosing words (for the sake of their weight rather than their grace), and in composing them, and in binding them in periods, there was nothing Antonius did not arrange in accord with the rules and, let us say, high art. He did this to an even greater degree, however, in the embellishment and shaping of the thoughts he expressed. Demosthenes, because he excels above all others in this category, has been judged by the learned to be the prince of orators. For what the Greeks call *schēmata* ["figures"] greatly embellish an orator and have weight not so much in verbal painting as in the illumination of expressed thoughts. Not only were these impressive in Antonius, but so too was his delivery exceptional. Let us divide the latter into gesture and voice. His gestures did not mark words but, rather, were in harmony with his thoughts: hands, shoulders, torso, the stamping of his foot, his step, and his whole movement, all agreeing with words and thoughts. His voice had endurance, but it was naturally hoarse (*vox permanens, verum subrauca natura*). Uniquely in his case, though, this defect was transformed into an asset. For in rhetorical laments, he had a certain tear-jerking quality (*flebile quiddam*), something which was well suited for both inspiring belief and producing pity, to such an extent that he offered confirmation of what is told of Demosthenes, who, asked what the most important thing was in oratory, replied, "Delivery," and then asked what the second most was, gave the same answer, and then the third, the same again. Nothing reaches more deeply into minds and molds, models, moves them—and makes orators seem to be the kind that they themselves want to be seen to be.[24]

In the middle of this remarkable passage is Cicero's revealing snapshot of the "exceptional delivery" of Antonius, comprising both "gesture" (*gestus*) and "voice" (*vox*). This is presented as supporting what follows: Demosthenes' assertion of the trebly supreme importance of "delivery" (*actio*), which more or less definitionally comprises those parts of an oratorical performance that cannot be transcribed. But in what comes just before, Cicero explains that Demosthenes himself has been judged supreme among orators primarily for his use of rhetorical figures, which of course can be written down and thus known to the "learned," that is, to later *readers* like Cicero himself. The eloquence of *this* Demosthenes is thus like that of "Antonius" (actually Cicero) in *On the Orator*, where, incidentally, the ideal orator of the title is largely theorized in terms not of *actio* but of *ornatus*—rhetorical "embellishment," such as the figures that made Demosthenes "prince of orators." *Actio, actio, actio* may make the live performer, but it is *ornatus, ornatus, ornatus* that fleshes out the recording star. In Cicero's case, this is true whether he is speaking under borrowed names in *On the Orator*, or as himself in the *Brutus*.

In the latter, a little while before the discussion of Antonius and Demosthenes, Cicero registers the rhetorical skill of the second-century commander and consul Servius Galba, who successfully defended himself from a charge of war crimes committed against a surrendering enemy. "What then is the reason," Cicero has Brutus ask, "if Galba possessed such talent as an orator, that none of this is evident in his speeches, a discrepancy at which I have no opportunity to marvel in the case of those who have left no written record at all?"[25] Cicero's answer begins with this latter category, which includes orators who already ignored the crucial role of writing in proper oratorical preparation (a matter on which Cicero had strong opinions, summarized here as "nothing profits speaking as much as writing") as well as those who, having delivered a well-received speech, thought better of messing with success by submitting it to further judgment in written form. Though Cicero initially insists that the unrecorded category is to be distinguished from that of the recorded but disappointing, they shade into one another as he turns to those whose motive not to leave a record is "that they believe they can speak better than they can write (*quod melius putent dicere se posse quam scribere*), which often happens to persons who are clever but insufficiently trained (*peringeniosis hominibus neque satis doctis*), as in the case of Galba, whom, we may say, a force not only of intellect but also of spirit, as well as a kind of instinctive pain, inflamed as he spoke and brought it about that his oratory was energetic, serious, and impassioned; but later, when he took up his pen while at rest and all that spirited energy, which people feel like a wind, had given out, his oratory drooped."[26] Cicero then offers a counterexample:

> Quod eis qui limatius dicendi consectantur genus accidere non solet, propterea quod prudentia numquam deficit oratorem, qua ille utens eodem modo possit et dicere et scribere; ardor animi non semper adest, isque cum consedit, omnis illa vis et quasi flamma oratoris exstinguitur. Hanc igitur ob causam videtur Laelii mens spirare etiam in scriptis, Galbae autem vis occidisse.
>
> This normally does not happen to those who have achieved a more polished kind of speaking, because here the orator is never failed by skill he may use both to speak and to write in the same way; warmth of spirit is not always present, and when it subsides all the force and, let us say, flame of the orator is extinguished. And so for this reason the mind of Laelius seems to me to breathe even in written form, while the force of Galba's perishes.[27]

In other words, in this "more polished kind of speaking" (*limatius genus dicendi*), exemplified by Gaius Laelius (whose father of the same name we met in Petrarch's *Africa*, and who will be Cicero's mouthpiece in his late dialogue *On Friendship*), unrecordable *actio* has entirely been eclipsed by what can instead be written down. The alleged motivation is the mitigation of flagging energy and enthusiasm during delivery, but the result is a more successful record. We may add that this is a record that is primed to be played, silently or aloud, by readers who likewise lack the "warmth of spirit" that the original occasion demanded.

We are so fond of remembering Cicero in action, in senate and Forum, that we have tended to overlook that his was a Laelian world of mediated voices as much as or more than it was an allegedly Demosthenic one of live *actio*. Even in his own lifetime, his oratorical reputation depended heavily on the publication of his orations, including of ones that were never delivered, like the bulk of the *Verrines*.[28] Then, of course, there is the whole literary career to which dialogues like *On the Orator* and the *Brutus* belong. In the prefaces to his later dialogues, Cicero often attempts to describe the relationship between his two modes as a zero-sum game: he found time to write when it was not possible to speak. But the two realms continually infect one another. What we want, he has Atticus say to him in the prologue to the *Brutus*, "is for you to write something (*ut aliquid . . . scribas*). To tell the truth, your writing fell silent a long while ago (*iampridem enim conticuerunt tuae litterae*). For we have heard (*accepimus*) absolutely nothing from you since you published (*edidisti*) *On the Republic* . . ."[29]

The *Brutus* was written under Caesar's dictatorship; *On Divination* was instead finished after the Ides of March, 44, and the prologue to its second book, after a résumé of the orator's literary output, promises a return to old priorities:

> In books I have been giving my opinion and addressing the people (*in libris enim sententiam dicebamus, contionabamur*), and I have been regarding philosophy as a substitute for my management of public affairs. But now, since I have begun to serve the republic with my counsel, my effort should be bestowed on the republic, or better, all thought and care should be placed in her, and only so much should be left for philosophical pursuits as will not be required for public duty and service.[30]

But this prediction would be proven wrong by Cicero's frenetic literary activity in the months to come, which saw the completion of five new treatises that would be his last: *On Fate*, *On Friendship*, *On Glory*, *Topics*, *On Duty*. In the very last of these, written while Cicero was writing the second and most celebrated of his *Philippics*, which was never meant to be delivered and which circulated instead in written form, the author seems to borrow a phrase from his own *Brutus*. When I realized, he writes, that "the republic no longer even existed" (the reference is notionally to the time of Caesar, but vague temporal markers make plain Cicero's doubt that the Ides of March had really changed anything), "the literature of Forum and senate naturally fell silent" (*illae scilicet litterae conticuerunt forenses et senatoriae*).[31] Nevertheless, the very phrase that Cicero seems to echo from the *Brutus*, *litterae conticuerunt*, reveals a fundamental shift in terms. There, Atticus was referring to the hiatus in Cicero's philosophical output since *On the Republic*, and his almost oxymoronic twinning of speaking and writing arguably only plays on the fiction that Cicero's treatises recorded real conversations: your written "dialogues" have stopped talking, Cicero. Here, however, the "silenced texts" are Cicero's speeches. Looking back over his career from what fate would make his final vantage point, Cicero sees only texts, and the alternatives of the *Brutus*, speak or write, have yielded fully to options internal to the latter: writing oratory versus writing philosophy. The fact is that, throughout his career, with or without a dictator, in triumph, in disgrace, or in restless limbo, in the *otium* of his seaside villas or in headlong flight into exile, Cicero could not stop writing, and in particular, writing things that sound (more and more) like Cicero. Imposed silence may occasionally have forced his hand (literally), but in the end, even he seems to recognize this as, simultaneously, a convenient accident that legitimized his full metamorphosis into a writer—and that of oratory into literature. For it was on the page, more than anywhere else, that he perfected that "more polished kind of speaking" that is actually a kind of writing, emblematized by the "file" (*lima*) that antiquity generally deployed as a metaphor (borrowed from carpentry and sculpture) for the manual labor not of orators, but of poets.

We can now appreciate that it is very much Cicero the Laelian who quotes Demosthenes on *actio* in the *Brutus*. Naturally, the great Greek is made to speak Latin (his triple answer would not have been *actio*, but *hupokrisis*[32])—and in indirect discourse. This Cicero carefully arranges in order to end not just rhythmically, but with a chiasmus: *quid secundum, idem, et idem tertium respondisse*. This, however, is child's play compared to the sentence that follows, bringing Crassus's excursus on Antonius to a close:

> Nulla res magis penetrat in animos eosque fingit format flectit, talesque oratores videri facit, quales ipsi se videri volunt.
>
> Nothing reaches more deeply into minds and molds, models, moves them—and makes orators seem to be the kind that they themselves want to be seen to be.[33]

The sentence's repeated consonants are easy to see: *fingit format flectit . . . ipsi se videri volunt*. But the ancient ear would have been even more pleased by the tricolon, that is, the sentence's arrangement into three parts, each of which Crassus punctuates rhythmically. The first closes without *clausula* but instead with the six long syllables that pound out the alliterated verbs (in asyndeton, the rhetorical figure of suppressing conjunctions) that themselves describe the hammering and shaping of the minds of listeners: ***fingit format flectit*** (––––––).[34] These six blows then soften into the *clausulae* that close the remaining two *cola*: first, a double cretic (*orato**res videri facit***: –⏑––⏑–), and then another (***se videri volunt***: –⏑––⏑–), a type that, after the infamous *esse videatur* (here deliberately winked at?), was among Cicero's favorites.[35] It all adds up to a display of rhetorical and rhythmical virtuosity that makes it clear not only that "Crassus" is really Cicero but also that the "prince of orators" may no longer be Demosthenes. This is, of course, yet another Oedipal victory; indeed, one could argue that the *Brutus* is nothing but a long line of vanquished father figures, from the distant Greeks to the mentors of Cicero's youth.

Cicero's most important victory, however, was over not his predecessors but himself. What, after all, did the *real* Cicero sound like? In his lifetime, even intimate friends like Atticus and Brutus, regular recipients of his letters and dedicatees of his dialogues, may sometimes have forgotten that there really was any Cicero but the written one, whom they were forever being called upon to serve as both readers and fictionalized interlocutors. In the *Brutus*, however, the written Cicero uncharacteristically consents to tell them (and, therefore, us) about the Cicero who cannot be read. This takes the form of a portrait of the orator as a young man, presented as a pendant to that of his slightly older rival, the great Hortensius, whose death in 50 BCE frames the entire dialogue:

> Nunc quoniam totum me non naevo aliquo aut crepundiis, sed corpore omni videris velle cognoscere, complectar non nulla etiam quae fortasse videantur minus necessaria. Erat eo tempore in nobis summa gracilitas et infirmitas corporis, procerum et tenue collum, qui habitus et quae figura non procul abesse putatur a vitae periculo, si accedit labor et laterum magna contentio; eoque magis hoc eos, quibus eram carus, commovebat, quod omnia sine remissione, sine varietate, vi summa vocis et totius corporis contentione dicebam. Itaque cum me et amici et medici hortarentur ut causas agere desisterem, quodvis potius periculum mihi adeundum quam a sperata dicendi gloria discedendum putavi. Sed cum censerem remissione et moderatione vocis et commutato genere dicendi me et periculum vitare posse et temperatius dicere, ut consuetudinem dicendi mutarem, ea causa mihi in Asiam proficiscendi fuit.

> Now, since you seem to want to know me not just from a particular mole or my baby rattle, but on the basis of my whole body, I shall embrace a few matters that may seem less important. Back then I had a body that was extremely thin and weak, with a long and slender neck—a physical makeup not thought to be remote from mortal risk, once one adds physical exertion and strain on the lungs. Further increasing the worry of those who cared about me was the fact that I pronounced everything without decreasing or varying my intensity, with maximum straining of my voice and whole body. Even so, when both friends and doctors urged me to stop trying cases, I thought any risk worth taking rather than giving up on success as an orator. But when I decided that, by relaxing and moderating my voice and by a change in oratorical style, I could simultaneously avoid mortal danger and speak in a more balanced way, this was the reason—to change my oratorical habits—that I left for the East.[36]

Having consulted teachers and trainers in Greece, Asia Minor, and Rhodes, Cicero at last resumed his barely begun career:

> Ita recepi me biennio post non modo exercitatior, sed prope mutatus; nam et contentio nimia vocis resederat et quasi deferverat oratio lateribusque vires et corpori mediocris habitus accesserat.
>
> Thus did I better myself in two years' time: not only more trained, but practically a changed man. For my excessive vocal strain had eased, my speaking style had cooled off, my lungs had gained strength, and my body, a bit of weight.[37]

In one sense Cicero does here show us what had happened to his "whole body": the scrawny beginner had come to look rather more like the well-fed pro we ourselves know from surviving portrait busts. But while we are invited to imagine that this represented a gain in vocal power (the obvious modern comparison being the opera singer), the transformed voice is itself described only in terms of what it has lost, namely, "strain" (*contentio*)—and this was not so much an innate characteristic as a mismatch of instrument and technique. Why did Cicero pass up the opportunity to describe his embodied voice to posterity and anyone else unable to hear him *viva voce*, especially in the enforced silence of a dictatorship he did not yet know he would outlive? Nowhere else in his treatises does he give himself a similar chance. It is as if his trip east had cured him not so much of a vocal defect as of the voice itself, neutralizing it, making it precisely unworthy of note. Indeed, comparison of his oratory before and after his trip has suggested that his teachers "restrained" his "excessive efforts" at the level of delivery because they "convinced him that such efforts were unnecessary and detrimental to his health and that his effects could be achieved no less by skillful management of his material than by sheer vocal effort, on which he had previously tended to rely."[38] This cure aimed to improve the young orator's delivery and save his health, but the shift in attention would have equally lasting side effects on the *written* Cicero.

Hortensius, by contrast, continued to rely extensively on what Cicero, in the same dialogue, tells us was "a pleasant, musical voice" (*vox canora et suavis*).[39] The phrase is ambiguous, pointing as much or more to his practiced delivery as to any innate timbre, though Cicero elsewhere lists the "sound of the voice" (*vocis sonus*) among the "natural gifts" with which an orator must be born.[40] In either case, the characterization is hardly the unqualified compliment it might at first seem to be, especially since it is immediately followed by the report that "his movement and gesture also possessed more artistry than was sufficient for an orator" (*motus et gestus etiam plus artis habebat quam erat oratori satis*). This is a milder version of the complaint that prompted a celebrated exchange between Hortensius himself and a critic, preserved for us by Aulus Gellius, who turns to Hortensius after describing the reputation of Demosthenes for unseemly attention to matters of dress and grooming:

> In the same way, Quintus Hortensius, more famous than any other orator of his age, with the exception of Marcus Tullius [Cicero], because he was dressed and draped with great elegance, with everything just in place, all around, and because while delivering his speeches his hands would be so very expressive and communicative, was pounded with insults and moral reproaches, and, even during legal proceedings and trials, many things were said against him that could have been said against an actor. But when Lucius Torquatus, a rather rustic and boorish fellow, said, with violent sarcasm, in front of the jury, when Sulla was on trial, that Hortensius was not merely an actor but a female mime, and called him "Dionysia," after the name of a notorious dancing girl, Hortensius replied, in a voice that was soft and gentle (*voce molli atque demissa*), "I, for one, would rather be a Dionysia—yes, a Dionysia—than what you are, Torquatus: devoid of the Muses, of Aphrodite, of Dionysus."[41]

Insults aside, Hortensius's voice helped to win him the adulation of his fans, but the publication of his orations yielded a paradox to which, long after his death, Quintilian would give pointed expression:

> ... eius scripta tantum infra famam sunt, qua diu princeps orator, aliquando aemulus Ciceronis existimatus est, novissime, quoad vixit, secundus, ut appareat placuisse aliquid eo dicente quod legentes non invenimus.
>
> His written texts fall so short of his reputation—according to which he was long judged the prince of orators, and then Cicero's rival, and finally, for the remainder of his life, second to Cicero—that it seems clear that people liked something when he spoke that we, as readers, do not find.[42]

In a bold reading of this and other passages, starting with Cicero's vocal self-portrait, Gualtiero Calboli finds in the eclipse of Hortensius the beginning of the end of the voice's centrality to classical oratory, not only in public delivery but also in private reading, destined to become ever more silent and solitary along a path that would lead, one day, to medieval monks and modern intellectuals.[43] But in addition to the fact that Quintilian himself provides us with one of antiquity's most extended treatments of delivery,[44] there is at least one very large step missing here. Cicero's age did indeed dethrone what Calboli calls the "orator without a microphone," but not in the sense, as he supposes, of one who needed nothing but his voice to reach and move a crowd. Rather, it required an orator who was ready to be miked for high-fidelity recording, something it found (at his own careful prompting) in Cicero himself. This, I would suggest, is what really lies at the heart of that age's supposed conflict, immortalized in Cicero's *Brutus*, between the oratorical exuberance he assigns to the newly coined rubric of "Asianism" and the opposed oratorical restraint of "Atticism." In the dialogue, Cicero makes Hortensius the exemplary Asianist and presents himself, in part to his own Atticist critics—such as his dedicatee—as occupying a compromise position on their side of the scale. Scholars have long disagreed about the extent to which the whole debate is a self-serving construction by Cicero himself, but its outcome is indisputable, and it hardly is (as Calboli slightly suggests) an Atticist victory that pairs increasingly plain style with decreasingly relevant delivery. Rather, Cicero himself wins both the battle and the war, defining the very reading habits that will lead Quintilian to find the written Hortensius painfully wanting. If Galba's unrecorded voice had offered Cicero a provocative puzzle, that of Hortensius provides Quintilian with a far more definitive silence made conspicuous by its historical proximity to what he can instead hear in abundance: his ideal orator, Cicero, "preeminent in all the aspects praised in other orators singly."[45]

"To me, speaking well and writing well seem to be one and the same" (*mihi unum atque idem videtur bene dicere ac bene scribere*), Quintilian confidently declares near the end of his twelve-book course on oratory, turning, promisingly, to the very question of the relationship between spoken and written orations.[46] But by this question he means only whether the former should diverge from plan when unexpected circumstances arise, and whether the latter should provide a verbatim "record of the oration as delivered" (*monumentum actionis habitae*). This leads to the following remarkable meditation:

> An Demosthenes male sic egisset ut scripsit, aut Cicero? Aut eos praestantissimos oratores alia re quam scriptis cognoscimus? Melius egerunt igitur an peius? Nam si peius, sic potius oportuit scribi ut dixerunt, si melius, sic potius oportuit scribi ut dixerunt. Quid ergo? Semper sic aget orator ut scribet? Si licebit, semper.
>
> Would Demosthenes, delivering a speech just as it was written, have delivered it badly? Or Cicero? Or do we know them as the most eminent of orators in any way other than from their written texts? Was their delivery better, therefore, or worse? If it was worse, then their speeches should have been delivered in the way that they wrote them; if better, then they should have been written in the way that they spoke. What then? Shall the orator give speeches just as he writes them? To the extent possible: always.[47]

Long forgotten here is the defeated voice of Hortensius—but so, too, are the unwritten voices of the victors, Demosthenes and Cicero, whom Quintilian is hardly alone in invoking as the twin paragons of the Greek and Latin rhetorical library. Cicero himself would claim Demosthenes as the primary Greek model of his own brand of high-volume "Atticism" and would borrow from his orations against Philip of Macedon an originally jesting title, which stuck, for his own final speeches against Mark Antony.[48] But Cicero no more knew Demosthenes' unwritten voice than did Quintilian know the voice of either.

"What if you had heard the beast himself?" So Aeschines is said to have asked the Rhodians after reading aloud, to great acclaim, the speech with which Demosthenes had ruined him.[49] The often repeated anecdote drove at least one ancient reader, Valerius Maximus, to exclaim, "Even if nothing can be added to his work, a large part *of* Demosthenes is nevertheless missing *in* Demosthenes, because he is read rather than heard."[50] It is actually the beginning part of this sentence that may be the most interesting, for it anticipates—and reverses—half of a more famous formulation by Quintilian, in a list of contrasts between the respective styles of Demosthenes and Cicero: "In the former, nothing can be taken away; in the latter, nothing added."[51] The comparison sounds clever (and is often quoted), but like the rest of Quintilian's list, it represents less a genuine antithesis than an effort to wrest difference from similarity. As already their respective positions at the top of the two languages of a shared rhetorical tradition would suggest, the two were hardly polar opposites, even if they were not quite the matched pair Cicero pretends they were to his skeptical Atticist friends. Quintilian, predictably, gives his beloved compatriot pride of place over his Greek counterpart, affirming that he matches the "power" (*vis*) of Demosthenes but adds the "richness" (*copia*) of Plato and the "charm" (*iucunditas*) of Isocrates.[52] But in this suggestion that Demosthenes lacked the other elements that added up to Cicero, do we find an echo of readers like Valerius, who found that what was really missing in the written Demosthenes was his voice? Quintilian, of course, had detected the same deficit in the written Hortensius. And most spectacularly of all, Cicero himself, amidst his praise of Demosthenes in reply to the Atticists, admits that even this greatest of Greek orators "does not satisfy me. Though he stands above all others in every kind of speaking, nevertheless, he does not always fill my ears, so greedy and capacious are they, always longing for something immense, infinite."[53]

Such gaps and disappointments, I would suggest, bracket the obscure object of desire of ancient rhetorical readers generally. Indeed, as James Porter observes, we are here very much in the realm of the sublime, including as theorized by Longinus, who, ironically, celebrates Cicero's own sublimity along with that of Demosthenes and who, Porter suggests, may even have in mind Cicero's own remarkable description of "eloquence which follows its course with roaring rapids, which all look up to, marvel at, and despair of achieving."[54] And this sublime object of hopeless admiration would seem, at first glance, to be the very voice that Lacanian critics, as we saw in the introduction, have identified as an endlessly alluring and elusive *objet petit a*.

Into this breach, however, boldly stepped Cicero, no less a pragmatist than Edison, who likewise found all earlier phonographs wanting. Cicero's intervention—not quite an invention, for it was mostly made of borrowed bits—was of a stylistic nature, but with a twist. Oratorical "style" (*lexis*) and oratorical "delivery" (*hupokrisis*) had been twinned problems already for Aristotle, who, in the *Rhetoric*, largely reduces the former to a principle of perspicuity to meaning and treats the latter only briefly, with painful reluctance.[55] His was an extreme position, but the basic imperative that sound be in the service of sense would prove persistent, at least at the level of theory. The orator was expected to fit his style (at least by choosing from the large readymade categories of the *genera dicendi*) to his setting and subject matter.[56] Likewise, the most common capsule principle of delivery was that the orator's voice should be fully expressive of his content, delivering cheerful things cheerfully, and so on, as Quintilian puts it in his discussion of what he flags as being of the "highest importance," namely *apta pronuntiatio*, delivery suited to sense.[57] It is largely on the basis of both principles that Dionysius of Halicarnassus, *pace* Valerius, declares that the sensitive and educated reader can readily recreate the delivery of Demosthenes by following the cues embedded in his shifting arguments and style, a matter on which the critic then provides an extended lesson.[58]

The appropriate delivery of the oratory of Isocrates, by contrast, Dionysius finds hard to deduce, a complaint that had repeatedly been addressed to this famous speechwriter (*logographos*) and teacher in his own lifetime, when a reportedly weak voice had led him to a career of guiding the speaking of others.[59] Isocrates, to be sure, had fought back, with the charge that his critics simply did not know how to read properly—indeed, that they sought to sabotage his reputation by deliberately reading his speeches badly.[60] The basic dilemma finds poignant if unwitting expression in Aristotle's fleeting discussion, in the *Rhetoric*, of what he calls the "written style" (*lexis graphikē*), which he opposes to a style that only works with the help of energetic delivery in the heat of the moment (*lexis agōnistikē*).[61] Arguing that "constant repetitions of words and phrases" should be avoided in the former, he brings home his point with a hypothetical example: "This is the villain among you who deceived you, who cheated you, who meant to betray you completely."[62] He thereby falls into his own trap, since we can only hear the monotony of its repetition if we can simultaneously hear what it would and should sound like, pronounced with feeling.

Further examples of a variously weighted tension between speaking, writing, and reading can be found across the whole ancient rhetorical tradition. It is against that backdrop that Cicero must have reacted to the hostile characterization of Hortensius as histrionic—by boors, the latter objected, who did not even understand the actor's art. The objection is key. Like his Muse-kissed colleague, Cicero recognized that the artistic voice could not entirely be reduced to its idoneity, but instead of following his lead in pursuit of polished delivery at the relative expense of style, he sought that voice in the paralinguistic sonorities of the latter. These, crucially, were writeable. Fully blurring the line between *lexis* and *hupokrisis*—well beyond anything yet attempted in Latin—Cicero redirected the "greedy and capacious" ears of readers to his own "Improvement in Phonographic or Speaking Machines" (to borrow the modest title of Edison's watershed patent). It is thus in Cicero's stylistic contrast with his two great oratorical rivals, one near (Hortensius) and one far (Demosthenes), that we find the fundamental mechanism of the vocal claim made for and by his written prose. The broader apparatus of that claim includes precisely the fact that, though he not only describes the celebrated sound of Hortensius but also repeats, three times, the story of Aeschines inadequately reading Demosthenes, he omits his own voice from the vocal portraits of the *Brutus* and other works, thus giving us no other Ciceronian voice to miss. Cicero's flesh-and-blood voice could hardly have been a bad one (this, surely, we would have been told by someone, probably cheerfully!), but unlike that of either of his rivals, his was made forgettable.[63] Whatever he himself actually sounded like, and however often he and his successors repeat, with stated approval, Demosthenic pronouncements on delivery, Cicero helped Roman rhetoric to lose a great deal of its unwriteable voice.

In part, as Calboli recognizes, what we have here is a divergence—hardly antiquity's first, as we have seen, but one with consequences stretching well beyond antiquity itself—between the performative and literary traditions of rhetoric. The effects would be felt almost as deeply in the continuing practice of the former as in the trajectories of the latter. Quintilian's enthusiastically Ciceronian oratorical course, generations later, makes this abundantly clear, rejecting the *vox canora et suavis* of Hortensius in favor of music that remains captivating in recorded form. "The student shall know he has made progress when he thoroughly enjoys Cicero" (*ille se profecisse sciat, cui Cicero valde placebit*), Quintilian famously proclaims, a phrase the aged Petrarch will later remember, in the letter we have several times considered, as the guiding principle of his own youth.[64] In other words, Petrarch was the distant heir to a fateful redirection of the Demosthenic imperative set in motion by Cicero himself. To find a voice, spoken or written, after Cicero, the *most* important thing would be to *read*: Cicero, Cicero, and Cicero.

Thus did Cicero succeed in replacing the "whole body" he so conspicuously does not give us in the *Brutus* with the vast, sonorous Ciceronian corpus by which he definitively became what he wanted "to be seen to be." To be sure, there originally was in all of this more than a slight strain of self-fashioning by the "new man" from Arpinum, addressing in Brutus, for example, someone thoroughly born to the cloth.[65] However, like the slight body that led him to overstrain an inadequate voice, Cicero's outsider status was a goad to what would prove to be, from the point of view of literary history, a far more important transformation. Like Petrarch, we know Cicero the way he wanted to be known: from his mole and his rattle, seeming tokens of a birthright, but which are better read as emblems for the signature marks and sounds he made for himself. In what only seems like a paradox, this signature also came to serve as a shibboleth of the republic of letters Cicero crafted in place of the one that no longer existed. The loss of the real republic, the loss of the real voice, the loss of the real Cicero: their sum is the Ciceronian real, which would come to be far more than a single voice.

Though almost immediately remembered as the day when "the mournful eloquence of the Latin tongue fell silent,"[66] Cicero's death definitively enabled his reanimation by those who would embrace his recorded voice as a paradigm. To be sure, beginning already in his own lifetime, there were anti-Ciceronians who sought to record differently, and from Sallust to Poliziano, many managed to offer something that was distinctly their own. The trouble, however, is that writing Latin that sounds either like or unlike Cicero is a bit like cutting a rock and roll record that sounds either like or unlike the Rolling Stones: one way or another, every band (a) sounds *something* like them while (b) inevitably demonstrating that they are *not* the Rolling Stones. More than defining a genre or even a language, Cicero too taught us what a record could sound like. Sounding less like Cicero would almost inevitably require making less sound; turn up the volume, and back he would come, never really having left. In other words, the only real way to account for Cicero's staggering influence on Latinity is to suppose that his achievement was both quantitatively and qualitatively technological, stretching his medium to its limits and then branding these as his own. It soon became almost impossible to write Latin prose without reproducing him: to the extent that readers heard a voice at all, it usually sounded something like Cicero.

One particularly powerful exception captured the attention of none other than Petrarch the reader and listener, and that was the voice of Augustine, whose *Confessions* he kept always with him and who would become his chief interlocutor in his final, private, posthumously circulated dialogue, generally called the *Secret*.[67] Stripping away everything he had learned about Latin composition from the resolutely Ciceronian curriculum of Late Antique education, Augustine produced a *vox clamantis* that is nothing short of shocking in its contrast. Prose rhythm is avoided obsessively.[68] Repetition is still everywhere, but it usually leaves words intact, repeating them whole, not in the parallel refrains of Ciceronian *anaphora*, but seriatim, earnestly pleading for understanding. Meaning is everything here, and in a passage that would long remain central to medieval debates about music and piety, the author seemingly rejects the word-bending song that had captivated him in his youth.[69] At the same time, as one scholar puts it, Augustine makes human words "sound strange or clumsy, inadequate to the task"; in our fallen condition, we must go on speaking, even jubilantly, but "in our own fumbling, repetitive, far from euphonious language"—in address to a God whose own condition is, by contrast, "silence and stillness."[70] But all of this finally leaves us with a question: Is the distinctiveness of Augustine's voice actually audible to someone who has not already read Cicero?

"Pick it up and read it," says a voice in the garden, leading Augustine to a nearby Bible and, definitively, to Christ.[71] This, however, is not the first scene of readerly conversion in the *Confessions*. Earlier, Augustine tells the God to whom the work is notionally addressed that his grammatical studies had brought him to "a book of a certain Cicero, whose tongue nearly everyone admired, but not so his heart. But that book contained that very man's exhortation to pursue philosophy and was titled *Hortensius*."[72] Lost to us, the *Hortensius* was, like the *Brutus*, a rare dialogue in which Cicero spoke as himself, though we know little about its contents. Augustine continues, addressing God, his notional reader, "And that book changed my disposition and changed my prayers directing them toward you, transforming my hopes and desires." He insists, "I did not apply myself to that book in order to sharpen my tongue, and I was persuaded not by how it spoke (*locutio*), but by what it was saying (*quod loquebatur*)."[73] Nevertheless, turning from Cicero to scripture, the latter "seemed to me unworthy of comparison to Ciceronian stateliness, for my inflated condition shrank from its measure (*modus*)," with this last word suggesting not just "moderation" but also "style" and "rhythm," or rather, scripture's lack thereof, at least to ears trained in a grammarian's classroom. Only later, in the garden, did Augustine really begin to read the word of God, with feeling. This was the necessary first step toward crafting the markedly un-Ciceronian voice exemplified by the *Confessions* themselves, a feat that would require cutting out "Cicero's tongue" at its root, replacing his Latin with one reshaped by and even stitched together from scripture, the strange sound of which already expresses its translator's commitment not to "how it spoke," but to "what it was saying." Augustine's contemporary Jerome would defend his Vulgate on precisely these grounds (and would be praised for such by Augustine himself), mindful that in the desert of his penance, to which he had dragged his library, God had accused him of being "not a Christian, but a Ciceronian."[74]

Nevertheless, precisely at the moment of his final conversion from sound to sense, Augustine hears rather more than two imperatives, telling him to "pick up" and "read." For one thing, he hears something like music, for the words are spoken *cum cantu*, by which he probably means that they are almost (but not quite) sung. Pronouncing the phrase, and "repeating it over and over," is a peculiar voice, "a boy's, or maybe a girl's—I am not sure." All of this ambiguity stands in for what is really in the balance here: whether this voice was human or divine.[75] But it infects the words themselves, which Augustine gives twice: *tolle lege, tolle lege*. The repeated singsong phrase combines rhyme (*tolle lege*) with consonance and assonance (*tolle lege*). Perhaps it is not surprising that God would sometimes sound like a child at play. At an even more basic level, the phrase offers a kind of heartbeat, in a work that invokes God as "the pulser of my ears, the enlightener of my heart," though in accord with Augustine's rejection of classical quantitative pronunciation, this trochaic pulse pointedly depends instead on stress accent: *tólle lége*.[76] But in the midst of his own rush to read (and write like) scripture, has Augustine unwittingly recorded God sounding not only like a broken record (which, as we saw in the last chapter, is the truest vocal record of all), but even just a little like Cicero, if only because his *Confessions* are so relentlessly noisy?

"What is this sound—so loud, so sweet—that fills my ears?" So one Scipio asks another as they hover high above the earth, in the famous dream sequence from the final book of Cicero's *On the Republic*, which would inspire in turn the first two books of Petrarch's *Africa*. This was the only part of the treatise to survive the end of antiquity—until the nineteenth century, when much of the rest was found lurking in an erased script over which had been written, in the sixth century, a copy of Augustine's *Commentary on the Psalms*, rigorously concerned with the what rather than the how of these ancient songs. What the younger Scipio hears is the harmony of the spheres, produced by the divinely orchestrated movement of the celestial bodies, supreme expression of the *logos* that governs all things, great and small. But as they gaze down at the republic below, their position is that of the treatise's own downward-looking readers, who likewise are surrounded by sounds both loud and sweet, emanating from *logos* in the ordinary sense of "language," that is, from the words they read, perhaps aloud. For Derrida, the former *logos* is just a retrojection of the latter, falsely naturalized and deliberately obscured by its assimilation, in Western philosophy, not to writing, but to speech. And a Derridean tale can certainly be told about what happens to this passage, preserved for us, even as the rest of the treatise went missing, by being englobed in a commentary by the late antique polymath Macrobius, endlessly interested in what "Scipio's Dream" can tell us about the order of the cosmos. But is Cicero's project really a philosophical one, here or elsewhere? His pursuit of the voice seems quite different from the "phono-logocentrism" attacked by Derrida, who, as we saw in our first chapter, insists on reading Aristotelian *phōnē* as "speech" rather than "voice." Cicero's phonocentrism is rather more like Edison's: much more interested in the *logos* as a means than as an end. Words are media here; so too is their music *and even their meaning*. What do these media express? Not an ineffable and unwriteable god but, rather, a deeply terrestrial "I am."

Ego enim idem sum. Inimici mei mea mihi, non me ipsum ademerunt. To be sure, these are words born of a desperate fear of erasure parlayed, in the writing that began once exile was over, into that most predictable of vanities: a bid for authorial immortality. But this did not prevent Petrarch from hearing in them (and in the rest of Cicero) a human presence, from believing that human artifacts do not simply represent or signify a human but are themselves as human as any other human body is, that they are, in Marshall McLuhan's well-worn phrase, "extensions of man."[77] But of which "man"? As we saw in the introduction, at least one modern philosopher, Adriana Cavarero, has found in the voice (of a woman singing) an arresting expression of difference: not just sexual difference, but the radical difference of each human being from every other. But even if plural bodies necessarily have plural voices, every seemingly new voice just as surely presents us with a sound we already know. This is plainer in the case of the recorded voice, which at its most singular emerges as the sound of a mass-produced machine: the *vox Ciceronis* becomes the voice of Latin rhetoric, the voice of classical Latin, and even, for Petrarch, the *vox humana* itself. "Ciceronian" voices—Cicero's for Latin prose, Petrarch's for Italian poetry—fundamentally offer only the mirage of radical identity in the recorded voice: not exceptions that prove the rule, but exceptionally definitive demonstrations of rules that they did not so much invent as infer. "Cicero," in other words, is not a freeform signature in endlessly pliant matter, but a map of mediatic possibilities and constraints. Truly Ciceronian users of that map (a category in which even the brilliant anti-Ciceronian Poliziano counted himself[78]) do not try to sound like Cicero; rather, they try to sound, like Cicero. That they usually wind up also sounding something like Cicero is a function of limits in the media they share with him. Those shared media include, at the most basic level, alphabetic writing itself, but they extend upward even into generic conventions, including ones that might at first seem banal, like the use of the first person: epistle, sonnet, and dialogue are all forever saying "I am."

Who then is the subject of this "I am"? Cicero? God? The record itself? The answer may not finally matter. Into the silence of the library at Verona, or wherever else we read, comes a voice. "Good morning! How do you do? How do you like the phonograph?": thus did Edison have his new machine address the assembled editors of *Scientific American*. To which Petrarch replies, "Francesco to his dear Cicero: Greetings. With enormous eagerness I have read through your letters..." As Michel de Certeau observes, "The voice makes people write."[79] But in the long phonographic age of which Edison is only a recent chapter, the voice has also been that which compels us to read.

Acknowledgments

Major portions of this book were written with the generous support of the Getty Research Institute, where I spent a year in idyllic surroundings as a Villa Scholar, with the further assistance of a University of California President's Fellowship in the Humanities. The remainder was written between the University of California, Los Angeles, and the University of Bristol, and during a few productive weeks at the American Academy in Rome. I would like to thank all of those institutions, especially their libraries.

As this project unfolded, drafts of this or that part were heard by the following audiences, all of whom are thanked, along with my hosts in each instance, for their questions and suggestions: the Center for Medieval and Renaissance Studies at UCLA; the Getty Villa; the London Latin Seminar at the Institute of Classical Studies; Yale University; the University of Liverpool; the Centre for Medieval Studies at the University of Bristol; the Congresso Internazionale di Studi Umanistici, Sassoferrato; the Max Planck Institute for the History of Science, Berlin; the Cambridge Philological Society; the Reception Seminar of the University of Oxford; the University of St. Andrews; the University of Reading; Durham University; "Libraries, Lives, and the Organization of Knowledge," a conference at the American Academy in Rome; and a brilliant group of graduate students who joined me for a seminar on the voice (and other sensory topics) at UCLA.

I have accumulated too many individual debts along the way to remember them all, so with apologies for any omissions, I would like to thank David Brafman, Caroline Bynum, Lorraine Daston, Michal Grover-Friedlander, Mary Hart, Joshua Katz, Laura Jansen, Ken Lapatin, Donka Minkova, Richard Pollard, Alex Purves, David Saunders, Alexa Sekyra, and Giulia Sissa. Special thanks to the friends and colleagues who read parts of the manuscript and offered countless corrections and improvements: Joshua Billings, Adam Lecznar, Emma Dillon, Sarah Nooter, Ellen O'Gorman, Mark Payne, Zrinka Stahuljak, and Mario Telò. Extra-special thanks to Johanna Drucker, Nina Eidsheim, and Brooke Holmes, who read and commented on the entire thing, and extra-extra-special thanks to Jim Porter, who read it twice and who has cheered me on from the beginning. Deep thanks also to Ramona Naddaff and Meighan Gale at Zone Books.

Finally, love and gratitude to Leonardo Proietti, who for ten years has been my voice of reason.

Notes

LINER NOTES

1. "The New Phonograph," *Scientific American Supplement* 632 (1888), 10096, celebrating the demonstration's tenth anniversary.

2. Friedrich A. Kittler, *Gramophone, Film, Typewriter*, trans. Geoffrey Winthrop-Young and Michael Wutz (Stanford: Stanford University Press, 1999), 21–114. So, too, Lisa Gitelman, *Scripts, Grooves, and Writing Machines: Representing Technology in the Edison Era* (Stanford: Stanford University Press, 1999). By contrast, Jonathan Sterne, *The Audible Past: Cultural Origins of Sound Reproduction* (Durham, NC: Duke University Press, 2003), professes "distaste for the cult of Edison in phonograph historiography" (29) and accordingly covers broader technological ground and reaches back, "archaeologically" (7), through the nineteenth and eighteenth centuries, since "many of the practices, ideas, and constructs associated with sound-reproduction technologies predated the machines themselves" (1). Interestingly, Sterne settles on a simple definition of "sound-reproduction technology" through its use of "*transducers*, which turn sound into something else and that something else back into sound" (22). It is not clear, however, how this definition fails to fit certain practices of writing and reading.

3. Marshall McLuhan, *The Gutenberg Galaxy: The Making of Typographic Man* (Toronto: University of Toronto Press, 1962).

4. "phonograph, n.," OED Online, http://www.oed.com/view/Entry/142654, accessed July 12, 2013.

5. *Grammē*, "line," would be a less obvious but arguably more suitable root, but the *OED* surmises that Berliner coined the term by inverting "phonogram," widely used since 1864 for the characters of Isaac Pitman's shorthand (also called

"phonographs") and already applied by some to phonograph records. Alphabetic writing was called "phonographic" (vs. "ideographic") as early as 1828; also influencing Edison's coinage (and the design of his machine) was Édouard-Léon Scott de Martinville's "phonautograph," which represented (but did not reproduce) sound vibrations. See the following entries in the OED Online: "gramophone, n.," http://www.oed.com/view/Entry/80603, "phonogram, n.," http://www.oed.com/view/Entry/142653, "phonographic, adj.," http://www.oed.com/view/Entry/142656, accessed July 12, 2013.

6. These last two opposed phrases are from Roy Harris, *The Origin of Writing* (London: Duckworth, 1986), 25–26. Harris's arguments are provocatively expanded and refined by David R. Olson, *The World on Paper: The Conceptual and Cognitive Implications of Writing and Reading* (Cambridge: Cambridge University Press, 1994), 65–90, with extensive references. For a brief introduction to writing's prehistory that is also very easy on the eyes, see Johanna Drucker and Emily McVarish, *Graphic! Design History* (Upper Saddle River, NJ: Pearson Prentice Hall, 2009), 2–27, abridged in the second (2013) edition.

7. On such stories, see Harris, *Origin*, 1–28.

8. On its evolution from other kinds of "phonographic" writing (versus "logographic" writing), see Olson, *World*, 78–84.

9. Calvert Watkins, *How to Kill a Dragon: Aspects of Indo-European Poetics* (New York: Oxford University Press, 1995), 23.

10. Barry B. Powell, "Homer and Writing," in *A New Companion to Homer*, ed. Ian Morris and Barry B. Powell (Leiden: Brill, 1997), 25.

11. Ibid.; a fuller version of his arguments can be found in Barry B. Powell, *Homer and the Origin of the Greek Alphabet* (Cambridge: Cambridge University Press, 1991).

12. "The vital connection between Homeric epic and the advent of the alphabet is essentially the opposite of that one promoted by Powell and his predecessors in this effort. The desire to write down Homer did not precipitate the creation of the Greek alphabet. Rather, the creation of the alphabet resulted in a writing down of Homeric verse": Roger D. Woodard, *Greek Writing from Knossos to Homer: A Linguistic Interpretation of the Origin of the Greek Alphabet and the Continuity of Ancient Greek Literacy* (New York: Oxford University Press, 1997), 256. See, further, Roger D. Woodard, "Phoinikēia Grammata," in *A Companion to the Ancient Greek Language*, ed. Egbert J. Bakker (Chichester, UK: Wiley-Blackwell, 2010), 25–46.

13. Notation and transcription reflect epic correption of the final vowel of πλάγχθη.

14. Paul Zumthor, *La poésie et la voix dans la civilisation médiévale* (Paris: Presses Universitaires de France, 1984), 11–12; my translation. Corrado Bologna, *Flatus vocis: Metafisica e antropologia della voce* (Bologna: Il Mulino, 1992), with a preface by Zumthor himself, offers a stimulating exploration of something similar across a broader range of material, much of it ancient.

15. Adriana Cavarero, *For More Than One Voice: Toward a Philosophy of Vocal Expression*, trans. Paul A. Kottman (Palo Alto: Stanford University Press, 2005), 7, 5. Compare Steven Connor, *Dumbstruck: A Cultural History of Ventriloquism* (Oxford: Oxford University Press, 2000), another key text in the burgeoning field of voice studies. Connor, however, takes the disembodied voice in a direction somewhat different from Cavarero's, finding in what he calls the "vocalic body" not so much the symptom of a single human body whose presence (and uniqueness) it stably guarantees as, instead, a force capable of precipitating a body even where there is none: "Voices are produced by bodies: but can also themselves produce bodies.... In bald actuality, it is we who assign voices to objects; phenomenologically, the fact that an unassigned voice must always imply a body means that it will always partly supply it as well" (35–36).

16. Cavarero, *For More Than One Voice*, 9.

17. Ibid., 13. Refining Levinas, Cavarero proposes, "In the realm of speech, it is not the face of the other, but rather his or her voice that constitutes the proper of uniqueness. And this voice, like the face but perhaps even more so, is 'neither sign which tends toward a signified, nor a mask that hides it.' Bound to a verbal system of signification—or, better, to logos—the voice is perfectly suited to the role of signifying the 'human fact' of uniqueness before and beyond this system" (28).

18. Anne Carson, "The Gender of Sound," in *Glass, Irony, and God* (New York: New Directions, 1995), 30.

19. Jean-Luc Nancy, *Listening*, trans. Charlotte Mandell (New York: Fordham University Press, 2007), 1. Roland Barthes, "Listening," in *The Responsibility of Forms: Critical Essays on Music, Art, and Representation*, trans. Richard Howard (New York: Hill and Wang, 1985), 245–60, likewise opens by distinguishing between the two words, albeit on different grounds.

20. Jacques Lacan, "Seminar 14 (20 April 1966): Summary of Crucial Problems; Jouissance," in *Seminar XIII: The Object of Psychoanalysis*, trans. Cormac

Gallagher, n.d., http://www.lacaninireland.com/web/published-works/seminars/, 151.

21. On can perhaps make a bit more out of Lacan's postulation of a fourth, "invocatory" drive (alongside the oral, scopic, and anal drives), which, in *The Four Fundamental Concepts of Psychoanalysis*, trans. Alan Sheridan, The Seminar of Jacques Lacan 11 (New York: W. W. Norton, 1981), he calls "the closest to the experience of the unconscious" (104), noting that it has "the privilege of not being able to close" (200).

22. Mladen Dolar, *A Voice and Nothing More* (Cambridge, MA: MIT Press, 2006), 31–32.

23. Cf. Michel de Certeau, "Vocal Utopias: Glossolalias," trans. Daniel Rosenberg, *Representations* 56 (1996), 34: "Savage voices and voices of the people, mad voices and infantile voices, define the places where it becomes possible and necessary to write. Voices furnish the hermeneutic with its condition of production, that is, with the sites it occupies where it converts them to text." See also Michel de Certeau, *The Practice of Everyday Life*, trans. Steven Rendall (Berkeley: University of California Press, 1984), 158.

24. Sigmund Freud, *New Introductory Lectures on Psycho-analysis*, trans. James Strachey (London: Vintage, 2001), 20.

25. Jacques Lacan, "The Instance of the Letter in the Unconscious, or Reason Since Freud," in *Écrits*, trans. Bruce Fink (New York: W. W. Norton, 2006), 413 [495].

26. William Empson, *Seven Types of Ambiguity*, 3rd ed. (Harmondsworth, UK: Penguin Books, 1961), 11.

27. Ibid., 10. Tellingly, Empson's quotation of Pope substitutes "be" for the original's less categorical "seem."

28. Ibid., 12.

29. Ibid., 13.

30. The poem thus bears, in linguistic and semiotic terms, a complexly "iconic" relationship to the soundscape it both describes and imitates. Similar relationships will occupy us often in the present book, which in this sense can be said to propose, inter alia, an investigation of the iconicity of literary language in relation to the voice as part of the (re)sounding world.

31. For brief introductions to Theocritean repetition in Tennyson, see Francis J. Hemelt, "Points of Resemblance in the Verse of Tennyson and Theocritus,"

Modern Language Notes 18.4 (1903): 115–17, and Wilfred P. Mustard, *Classical Echoes in Tennyson* (New York: Macmillan, 1904), 31–43.

32. John M. Picker, *Victorian Soundscapes* (Oxford: Oxford University Press, 2003), 123–26, with bibliography and excellent discussion. The recording of "The Charge of the Light Brigade" can be heard here: http://en.wikipedia.org/wiki/File:LightBrigade-Tennyson.ogg.

33. W.H. Preece, *The Phonograph; or, Speaking and Singing Machine, Invented and Patented by Thomas Alva Edison, with Extracts from the Principal Journals in England and America* (London: London Stereoscopic, 1878), 52, which I have been unable to obtain and quote instead from Picker, *Victorian Soundscapes*, 187 n. 30.

34. "In place of written or spoken verbal expression, 'Break, Break, Break' offers sound, specifically the breaking of waves on stones," observes Jason R. Rudy, *Electric Meters: Victorian Physiological Poetics* (Athens: Ohio University Press, 2009), 72, offering a suggestive reading of the poem less as an elegy than as a meditation on the ends and limits of lyric in which the "voice that is still," like the "vanished hand" in the line before, is as much the poet's as it is his dead friend's; especially intriguing is Rudy's observation of "the rhythmic superfluity of the opening stanza's 'I'" (ibid.).

35. Tennyson, "The Miller's Daughter."

36. Florence Dupont, *The Invention of Literature: From Greek Intoxication to the Latin Book*, trans. Janet Lloyd (Baltimore: Johns Hopkins University Press, 1999), 22.

37. Maurizio Bettini, *Voci: Antropologia sonora del mondo antico* (Turin: G. Einaudi, 2008). See also Giovanni Manetti, "Semanticità, articolazione, scrivibilità: Gli spazi di confine tra l'uomo e l'animale nella Grecia antica," in *Ta zōia. L'espai a Grècia II: Els animals i l'espai*, ed. Montserrat Jufresa and Montserrat Reig (Tarragona: Institut Català d'Arqueologica Clàssica, 2011), 13–26. Manetti carefully assesses the relationship between "continuity" and "discontinuity" between humans and animals in, especially, Aristotelian theories of voice and language. Bettini pays special attention to birdsong, and for an analogous effort to place medieval musicality against such a backdrop, see Elizabeth Eva Leach, *Sung Birds: Music, Nature, and Poetry in the Later Middle Ages* (Ithaca, NY: Cornell University Press, 2007).

38. Mark Payne, *The Animal Part: Human and Other Animals in the Poetic Imagination* (Chicago: University of Chicago Press, 2010); also "The Understanding Ear: Synaesthesia, Paraesthesia and Talking Animals," in *Synaesthesia and the Ancient*

Senses, ed. Shane Butler and Alex Purves (Durham, UK: Acumen, 2013), 43–52.

39. Susan Stewart, *Poetry and the Fate of the Senses* (Chicago: University of Chicago Press, 2002), 109–10, which is followed by a meditation, in Stewart's own inimitable voice, on the shared, social nature of vocal individuality: "The 'object' of my love for your voice emerges in the relation between my history and the uniqueness of your existence, the particular timbre, tone, hesitations, and features of articulation by which all the voices subject to your own history have shaped your voice's instrument. In listening, I am listening to the material history of your connection to all the dead and the living who have been impressed upon you. The voice, with the eyes, holds within itself the life of the self—it cannot be another's" (110).

40. Stefan Hagel, *Ancient Greek Music: A New Technical History* (Cambridge: Cambridge University Press, 2010), ambitiously supplementing and revising M. L. West, *Ancient Greek Music* (Oxford: Oxford University Press, 1992), which however remains the indispensible technical guide. Thomas J. Mathiesen, *Apollo's Lyre: Greek Music and Music Theory in Antiquity and the Middle Ages* (Lincoln: University of Nebraska Press, 1999), 1–19, offers a readable survey of the kinds of evidence available and the history of scholarship on the subject.

TRACK ONE: BODY AND SOUL

1. Pollux, *Onomasticon* 9.1. A brief introduction to the work and its organization is given by Eleanor Dickey, *Ancient Greek Scholarship* (New York: Oxford University Press, 2006), 96.

2. Pollux, *Onomasticon* 9.51.

3. Giuseppe Zecchini, "Polluce e la politica culturale di Commodo," in *L'Onomasticon di Giulio Polluce: Tra lessicografia e antiquaria*, ed. Cinzia Bearzot, Franca Landucci Gattinoni, and Giuseppe Zecchini (Milan: Vita e Pensiero, 2007), 17–26.

4. There is an excellent discussion in Graham Anderson, *Lucian: Theme and Variation in the Second Sophistic* (Leiden: Brill Archive, 1976), 67–71.

5. Lucian, *A Professor of Public Speaking (Rhetorum praeceptor)* 11, quoted here from the Loeb edition of Lucian, vol. 4, trans. A.M. Harmon (London: Heinemann, 1913).

6. Ibid. 24.

7. Homer, *Iliad* 1.249. Interestingly, "honeyed" does not appear among the list

of vocal epithets provided by Pollux himself in *Onomasticon* 2.114–17.

8. Philostratus's account of Pollux is in *Lives of the Sophists* 2.12.592–3. The "honeyed voice" appears near the end of the chapter.

9. Cf. *dialexis*, and on both see Karl-Heinz Uthemann, "Diatribe (A)," in *Brill's New Pauly*, ed. Hubert Cancik and Helmuth Schneider (Brill Online, 2012), http://referenceworks.brillonline.com/entries/brill-s-new-pauly/diatribe-e316870.

10. Philostratus, *Lives of the Sophists* 2.12.593. This and the following translation are borrowed from the Loeb edition of Philostratus and Eunapius, *Lives of the Sophists*, trans. Wilmer Cave Wright (London: Heinemann, 1922).

11. Ibid. Wright takes the preceding sentence to be Pollux's introduction of the letter and thus part of Philostratus's quotation, but it looks rather more like a parenthesis by the latter or an interpolated gloss.

12. Thomas Schmitz, "Narrator and Audience in Philostratus' *Lives of the Sophists*," in *Philostratus*, ed. Ewen L. Bowie and Jaś Elsner (Cambridge: Cambridge University Press, 2009), 52–53: "Rhetorical handbooks taught when and how orators should pretend to be improvising (when they had really prepared their speeches in advance and in writing). It is thus not surprising that Philostratus' narrator should in his turn be imitating an oral performance. While an oral speech would present its author in the flesh, Philostratus had to use textual markers to achieve his presence; his powerful and often emotional interventions should be seen in this context."

13. Philostratus, *Lives of the Sophists*, preface, alluding to Homer, *Odyssey* 4.220–21.

14. For other examples of Philostratean attention to vocal quality, see Francesco De Martino, "La voce degli autori," in *Lo spettacolo delle voci*, ed. Francesco De Martino and Alan H. Sommerstein (Bari: Levante, 1995), 32–33, and (with special attention to metaphors like that used to describe Pollux) Graham Anderson, *Philostratus: Biography and Belles Lettres in the Second Century A.D.* (London: Croom Helm, 1986), 36, who notes, "The art of eloquence had to be eloquently described, and accordingly flood-tides of it pour out from every page . . ."

15. The connection is further helped by the verbal similarity of *hēdonē* ("pleasure") to *hēdus* ("sweet").

16. On other Greek words for "voice," some of which, such as *audē*, preferred by Homer, also double as "speech," see Guy Lachenaud, *Les routes de la voix: L'antiquité grecque et le mystère de la voix* (Paris: Les Belles Lettres, 2013), 37–44.

17. Edward H. Johnson, "A Wonderful Invention: Speech Capable of Indefinite Repetition from Automatic Records," *Scientific American* 37.20 (November 17, 1877): 304. "If there was a defining figure in early accounts of sound recording, it was the possibility of preserving the voice beyond the death of the speaker," notes Sterne, *Audible Past*, 287.

18. Here I follow Calvert Watkins, *The American Heritage Dictionary of Indo-European Roots*, 3rd ed. (Boston: Houghton Mifflin Harcourt, 2011), 7, 100. R.S.P. Beekes, *Etymological Dictionary of Greek* (Leiden: Brill, 2010), discusses (but rejects) objections to the derivation of *phōnē* and its relation to *phēmi* on the basis of a difference between sounding and saying. Lachenaud, *Routes de la voix*, 15, claims a voice/speech distinction between **wek*^{w}- and **wer(dh)*^{w}- (Watkins **werə-*^{3}), but only on the selective basis of historical forms (like our "word") that depend on them.

19. For the correct view, see Bernard M. W. Knox, "Silent Reading in Antiquity," *Greek, Roman, and Byzantine Studies* 9.4 (1968): 421–35; M.F. Burnyeat, "Postscript on Silent Reading," *Classical Quarterly* 47.1 (1997): 74–76; A.K. Gavrilov, "Techniques of Reading in Classical Antiquity," *Classical Quarterly*, 47.1 (1997): 56–73; William A. Johnson, "Toward a Sociology of Reading in Classical Antiquity," *American Journal of Philology* 121.4 (2000): 593–627, and, revising and expanding this, *Readers and Reading Culture in the High Roman Empire: A Study of Elite Communities* (New York: Oxford University Press, 2010), 3–16. Johnson in particular offers a constructive look beyond the narrow question of silent reading itself. The inaccurate view that silent reading in antiquity was instead so rare as to be startling leans heavily on Augustine's description of Ambrose reading silently (*Confessions* 6.3), but, in an article that should now be added to the subject's essential bibliography, William North, "Hearing Voices in Late Antiquity: An Aural Approach to Augustine's Confessions," in *The Middle Ages in Texts and Texture: Reflections on Medieval Sources* (Toronto: University of Toronto Press, 2011), 17–18, offers the persuasive suggestion that the passage should primarily be read as an allegorical prefiguration of the stunning silence, broken only by the voice of God, that is the putative endpoint of Christian history. On trained "readers" (Greek *anagnōstēs*, Latin *lectores*), see especially Johnson, "Toward a Sociology," 619–21; Raymond J. Starr, "Reading Aloud: *Lectores* and Roman Reading," *Classical Journal* 86.4 (1991): 337–43.

20. Jesper Svenbro, "The 'Interior' Voice: On the Invention of Silent Reading," in *Nothing to Do with Dionysos? Athenian Drama in Its Social Context*, ed. John

J. Winkler and Froma I. Zeitlin (Princeton: Princeton University Press, 1990), 369, 372.

21. Lucretius, *On the Nature of Things* 4.590–91.

22. Jacques Derrida, *Of Grammatology*, trans. Gayatri Chakravorty Spivak (Baltimore: Johns Hopkins University Press, 1976), 11.

23. Analogous to the Aristotelian misreading to be described below is Derrida's misrepresentation of Saussure on the voice, on which see James I. Porter, "Saussure and Derrida on the Figure of the Voice," *MLN* 101.4 (1986): 871–94.

24. Aristotle, *Organon. I. Catégories. II. De l'interprétation*, trans. J. Tricot (Paris: Vrin, 1946), 77–78. Translation reproduced (without attribution) in Jacques Derrida, *De la grammatologie* (Paris: Éditions de Minuit, 1967), 46.

25. In Aristotle, *The Works of Aristotle*, ed. W.D. Ross (Oxford: Oxford University Press, 1928), 1:25. Translation reproduced (also without attribution) in Derrida, *Of Grammatology*, 11.

26. Scholars have tended to be as farsighted as translators here, seeking to connect the sentence to the rest of the treatise and to the broader project of Aristotelian thought about language and meaning before really sorting out its peculiar wording. Our focus will instead be on metaphor and anatomy, but for an extensive study of the sentence in its ordinary philosophical context, see Deborah K. W. Modrak, *Aristotle's Theory of Language and Meaning* (Cambridge: Cambridge University Press, 2001), with earlier bibliography.

27. Plato, *Theaetetus* 192a–195a.

28. Aristotle, *On Memory* 1.450a (trans. J.I. Beare). On cited translations of Aristotle in this book, see the note at the head of the bibliography.

29. Indeed, Aristotle (attributed), *On Things Heard* 801b, applies the same Platonic metaphor of sealed wax to the voice itself: "Voices are distinct in proportion to the accuracy of the sounds uttered; for it is impossible for the voice to be distinct if the sounds are not perfectly articulated, just as the sealing of signet-rings cannot be distinct unless they are accurately impressed" (trans. T. Loveday and E. S. Forster). *On Things Heard* is largely about the kinds of sounds made by human voices; on this crucial treatise and its authorship, with a translation and notes, see Andrew Barker, ed., *Greek Musical Writings* (Cambridge: Cambridge University Press, 1989), 2:98–109.

30. Jacques Derrida, "Freud and the Scene of Writing," in *Writing and Difference*, trans. Alan Bass (Chicago: University of Chicago Press, 1978), 197.

31. Liddell and Scott, s. v. σύμβολον. Much more in Walter Müri, "ΣΥΜΒΟΛΟΝ: Wort- und sachgeschichtliche Studie," in *Griechische Studien: Ausgewählte Wort und sachgeschichtliche Forschungen zur Antike* (Basel: Friedrich Reinhardt, 1976), 1–44.

32. Peter Struck, *Birth of the Symbol: Ancient Readers at the Limits of Their Texts* (Princeton: Princeton University Press, 2004), 84. Norman Kretzmann, "Aristotle on Spoken Sound Significant by Convention," in *Ancient Logic and Its Modern Interpretations*, ed. John Corcoran (Dordrecht: Reidel, 1974), 3–21, takes *sumbolon* to mean "symbol" but helpfully distinguishes it from *sēmeion* ("sign" or "symptom"), undoing damage that, as he notes, is as least as old as Boethius's translation of both as *nota* (9, 18–19 n. 6, with a brief survey of subsequent translations); his careful unpacking of the passage in context leads him to the conclusion that its real emphasis is on *kata sunthēkēn*, which we shall discuss, and that Aristotle's "purpose is not to do psychology or epistemology but rather to provide grounds for the conventionalism he is going to proclaim" (11).

33. Pollux, *Onomasticon* 9.70.

34. On the general question of the ancient interplay between verbal and monetary symbolic systems, see Marc Shell, *The Economy of Literature* (Baltimore: Johns Hopkins University Press, 1978).

35. Plato, *Symposium* 191d. Aristotle himself describes lovers striving for one another like two *sumbola* at *Eudemian Ethics* 1239b.

36. Aristotle, *On the Soul* 2.2.414a (trans. J. A. Smith).

37. Ibid. 2.1.412b (trans. J. A. Smith). On this passage, and by way of introduction to the sometimes fierce debates about hylomorphism and Aristotle's theory of mind, see Martha C. Nussbaum and Hilary Putnam, "Changing Aristotle's Mind," in *Essays on Aristotle's "De Anima,"* ed. Martha C. Nussbaum and Amélie Oksenberg Rorty (Oxford: Oxford University Press, 1992), 45–46. Also Philip J. van der Eijk, "Aristotle's Psycho-Physiological Account of the Soul-Body Relationship," in *Psyche and Soma: Physicians and Metaphysicians on the Mind-Body Problem from Antiquity to Enlightenment*, ed. John P. Wright and Paul Potter (Oxford: Oxford University Press, 2000), 57–77.

38. Aristotle, *On the Soul* 2.8.420b (trans. J. A. Smith). Steven Connor's *Beyond Words: Sobs, Hums, Stutters and Other Vocalizations* (London: Reaktion Books, 2014) amusingly and provocatively opens with Aristotle's troublesome cough.

39. Aristotle, *History of Animals* 4.9.535a–b (trans. D. W. Thompson). There is a brief but perceptive discussion of this and other passages in which Aristotle

probes the relationship between human expression (and expressive anatomy) and that of other animals in Mark Payne, *The Animal Part: Human and Other Animals in the Poetic Imagination* (Chicago: University of Chicago Press, 2010), 85–88. Fuller discussion of the three categories here (*psophos*, *phōnē*, and *dialektos*) in Giovanni Manetti, "Semanticità, articolazione, scrivibilità: Gli spazi di confine tra l'uomo e l'animale nella Grecia antica," in *Ta zōia. L'espai a Grècia II: Els animals i l'espai*, ed. Montserrat Jufresa and Montserrat Reig (Tarragona: Institut Català d'Arqueologica Clàssica, 2011), 13–14. Anatomical distinctions similar to Aristotle's are made by Galen; relevant passages are collected by Jeffrey L. Wollock, *The Noblest Animate Motion: Speech, Physiology, and Medicine in Pre-Cartesian Linguistic Thought* (Amsterdam: John Benjamins, 1997), 11–12. For Francesco De Martino, "La voce degli autori," in *Lo spettacolo delle voci*, ed. Francesco De Martino and Alan H. Sommerstein (Bari: Levante, 1995), 19–21, a definitive (anatomical) distinction between (articulate) "speech" (as *dialektos*) and "voice" (as *phōnē*) comes only in Galen; he helpfully provides (n. 4) an extensive list of both lost and surviving ancient theoretical discussions of the matter.

40. Aristotle, *On Youth, Old Age, Life and Death, and Respiration* 4.469b.

41. Aristotle, *History of Animals* 1.17.496a (trans. D. W. Thompson).

42. Ibid. 1.16.495b (trans. D. W. Thompson). I take Aristotle no longer to be speaking solely of ovipara at this point.

43. Aristotle, *On the Soul* 2.8.420b; 3.3.427a–429a.

44. The South Italian Greek group is sometimes titled *Poet as Orpheus with Two Sirens*, though it is not certain that the seated figure alludes to Orpheus; its current arrangement (with the Sirens flanking the poet) is conjectural. Partial discussion in Angelo Bottini and Pier Giovanni Guzzo, "Orfeo e le Sirene al Getty Museum," *Ostraka: Rivista di antichità* 2.1 (1993): 43–52. Full discussion, with a proposed dating to the final decades of the fourth century, will appear in the online catalogue by Maria Lucia Ferruzza, *Ancient Terracottas from South Italy and Sicily in the J. Paul Getty Museum* (Malibu, CA: J. Paul Getty Museum, forthcoming). Though she does not discuss the group, Susan Stewart, *Poetry and the Fate of the Senses* (Chicago: University of Chicago Press, 2002), 58, 106, uses images of the poet and of one of the Sirens as frontispieces to her chapters on, respectively, "Sound" and "Voice and Possession."

45. Unless, of course, we are listening for the wrong thing, which, regarding the haughty hero of the *Odyssey*, is the wry suggestion of Franz Kafka, "The

Silence of the Sirens," in *Homer: A Collection of Critical Essays*, ed. George Steiner and Robert Fagles (Englewood Cliffs, NJ: Prentice-Hall, 1962), 98: "But Ulysses, if one may so express it, did not hear their silence; he thought they were singing and that he alone did not hear them. For a fleeting moment he saw their throats rising and falling, their breasts lifting, their eyes filled with tears, their lips half-parted, but believed that these were accompaniments to the airs which died unheard around him."

46. For another reading, complementary to the one I have attempted here, of Derrida's complex misreading of the opening of *On Interpretation*, see Joseph Claude Evans, *Strategies of Deconstruction: Derrida and the Myth of the Voice* (Minneapolis: University of Minnesota Press, 1991), 147–51. In a different vein, Veit Erlmann, *Reason and Resonance: A History of Modern Aurality* (New York: Zone Books, 2010), 11–12, 47–49, 60, queries the role that the tympan, the functioning of which Derrida misunderstands, plays in his critique of the "metaphysics of presence."

47. The same intuitive anatomical link between *psukhē* and *phōnē* underpins polemic about the inauthentic voices of actors and duplicitous orators, such as that of Demosthenes against Aeschines, on which see Pat Easterling, "Actors and Voices: Reading Between the Lines in Aeschines and Demosthenes," in *Performance Culture and Athenian Democracy*, ed. Simon Goldhill and Robin Osborne (Cambridge: Cambridge University Press, 1999), 159. The extent to which the role(s) of the voice in ancient culture shape(s) anatomies of the soul and mind clearly deserves rather more attention than it seems to have received.

48. Aristotle, *On Interpretation* 2.16a (trans. J. L. Ackrill, modified).

49. Ibid. 1.16a (trans. J. L. Ackrill).

50. Or τῷ πάθει μάθος ("we learn by suffering"), in the slightly less epigrammatic version given by Aeschylus, *Agamemnon* 176, where the main meaning of *pathos* is "suffering," though the proverb accesses the word's broader meaning.

51. Aristotle, *On Interpretation* 4.16b (in my own partial translation).

52. Aristotle, *On the Soul* 2.8.420b.

53. Ibid. (trans. J. A. Smith).

54. Aristotle, *History of Animals* 4.9.535a. On *aphōna*, see also *Poetics* 20.1456b.

55. Gorgias, *Encomium of Helen* 19.

56. Aristotle, *Rhetoric* 3.1.1404a; *Poetics* 6.1450b.

57. Oscar Wilde, "The Ballad of Reading Gaol," in *The Complete Works of Oscar Wilde*, ed. Bobby Fong and Karl Beckson (Oxford: Oxford University Press,

2000), 1:196.

58. "The Phonograph," *Harper's Weekly*, March 30, 1878: 249d. Similar feats (including the transmission of coughs) were celebrated for the telephone, on which see Steven Connor, *Dumbstruck: A Cultural History of Ventriloquism* (Oxford: Oxford University Press, 2000), 380–82.

59. Thomas A. Edison, "Improvement in Phonograph or Speaking Machine," February 19, 1878, http://edison.rutgers.edu/patent1.htm.

60. Friedrich A. Kittler, *Gramophone, Film, Typewriter*, trans. Geoffrey Winthrop-Young and Michael Wutz (Stanford: Stanford University Press, 1999), 21–22.

61. Perhaps, however, coughs and other such sounds are distractions from a more important vanishing point of phonographic technologies and desires: silence. For a rich exploration of silence in ancient (Greek) culture, see Silvia Montiglio, *Silence in the Land of Logos* (Princeton: Princeton University Press, 2000). "Emphasizing verbally an act of silence is a pervasive pattern in Greek literature" (289): Montiglio is largely concerned with somewhat extended silences, that is, with silence as speech's interruption or suppression, and so, in literature, with descriptions of silence rather than its direct representation. She generally reads such silences as negative, against a cultural backdrop in which speech and power were inseparable, unless silence could itself be made into a kind of speech or action, as in the preteritions of a skilled orator or the quiet plotting of Odysseus (both of which, in any case, are only temporary). But the literary material she collects makes me wonder whether, sometimes, the silence-theme may not also represent an amplification of something far smaller, namely, the sound-breaks that are essential to speech itself and the pauses, however brief, that are no less a part of a vocalization (and of vocal aesthetics) than sound is. "For a poet, losing his voice would mean to die as a poet" (290): but in another sense (at each caesura, etc.), the poet, and the written poem, must continually fall silent.

62. More broadly on the way in which sound recording constructs both original and copy, see Jonathan Sterne, *The Audible Past: Cultural Origins of Sound Reproduction* (Durham, NC: Duke University Press, 2003), 214–86: "This is the salient point for all reproduction: it is not just eavesdropping on live performance; it is a studio art" (237).

63. Contrast, however, Julia Kristeva on the (poetic) "genotext," to which we shall turn in the next chapter, and again, in Chapter 3, via Roland Barthes.

TRACK TWO: FALLING IN LOVE AGAIN

1. Jacques Lacan, "The Mirror Stage as Formative of the *I* Function as Revealed in Psychoanalytic Experience," in *Écrits*, trans. Bruce Fink (New York: Norton, 2006), 75–81, first published in 1949.

2. George Peter Murdock, "Cross-Language Parallels in Parental Kin Terms," *Anthropological Linguistics* 1.9 (1959): 1–5. On Murdock's ties to Hoover's FBI, see David H. Price, *Threatening Anthropology: McCarthyism and the FBI's Surveillance of Activist Anthropologists* (Durham, NC: Duke University Press, 2004), 70–89.

3. Roman Jakobson, "Why 'Mama' and 'Papa'?" in *Selected Writings*, ed. Stephen Rudy (The Hague: Mouton, 1962), 538–45. On Greek and Latin nursery words (which follow the pattern), see Mark Golden, "Baby Talk and Child Language in Ancient Greece," in *Lo spettacolo delle voci*, ed. Francesco De Martino and Alan H. Sommerstein (Bari: Levante, 1995), 11–33, with earlier bibliography, and Wilhelm Heraeus, "Die Sprache der römische Kinderstube," *Archiv für lateinische Lexikographie und Grammatik* 13 (1904): 149–72, who partially anticipates (149–50) Jakobson's conclusions.

4. Some minor revisions to Jakobson's account have been proposed; for a brief summary, see Trevor A. Harley, *The Psychology of Language: From Data to Theory*, 3rd ed. (New York: Psychology Press, 2008), 123–24. Daniel Heller-Roazen, *Echolalias: On the Forgetting of Language* (New York: Zone Books, 2005) begins his own investigation of the limits of speech (and of the various ways in which speech can be gained and lost) with Jakobson and "the indistinct and immemorial babble that, in being lost, allowed all languages to be" (12).

5. Jakobson, "Why," 541–42, interestingly preferring the paternal word.

6. The joke was on the Greeks: the word "was ironically oriental in origin" (cf. Babylonian-Sumerian *barbaru*), as notes Edith Hall, *Inventing the Barbarian: Greek Self-Definition Through Tragedy* (Oxford: Oxford University Press, 1989), 4, with discussion of the history of the term and its persistently linguistic resonance, 4–12 and passim.

7. Aristophanes, *Frogs* 209ff.

8. Andrew L. Ford, "Dionysos' Many Names in Aristophanes' Frogs," in *A Different God? Dionysos and Ancient Polytheism*, ed. Renate Schlesier (Berlin: De Gruyter, 2012), 343–55.

9. Statius, *Thebaid* 5.613–14.

10. Lynn Enterline, *The Rhetoric of the Body from Ovid to Shakespeare*

(Cambridge: Cambridge University Press, 2000), 1, 38, reading Ovid, *Metamorphoses* 6.558. "I call Ovid's trope of the voice 'phonographic' because the kinds of self-endorsing fantasies that Derrida describes as 'phonocentric' are no sooner entertained in the *Metamorphoses* than they are eroded" (12): Enterline's exploration of the productive crisis posed for the (authorial) subject by "the sound a voice makes when it fails to work" (18), which she pursues, with equal insight, in both aesthetic and erotic terms, provides one of the most interesting and compelling works of Ovidian scholarship out there; along the way, she acutely detects a series of the poem's extended figures for the voice, to her unpacking of which the reader of the present study is enthusiastically referred.

11. Ovid, *Metamorphoses* 10.383–84. Also noted by Victoria Rimell, *Ovid's Lovers: Desire, Difference and the Poetic Imagination* (Cambridge: Cambridge University Press, 2006), 119, who compares Philomela's tongue and, as will I, Orpheus (who is narrating Myrrha's story).

12. Ovid, *Metamorphoses* 1.206; cf. 9.421.

13. Ibid. 10.702, 14.280–81. Susan Stewart, *Poetry and the Fate of the Senses* (Chicago: University of Chicago Press, 2002), 60–61, launches a chapter on "Sound" with a set of similarly nonlinguistic sounds made during and after Ovidian metamorphosis, including those of the Actaeon episode, to which we shall turn in a moment.

14. Ovid, *Metamorphoses* 12.49–51: *nec tamen est clamor, sed parvae murmura vocis, / qualia de pelagi, si quis procul audiat, undis / esse solent.*

15. Ibid. 2.455–56; 5.587.

16. *Thesaurus Linguae Latinae*, "murmur" 8.1675.51–1676.37.

17. Heraclitus, frag. 91 (ed. Diels-Kranz).

18. Ovid, *Metamorphoses* 11.50–53.

19. Many ancient writers describe the painting (or copies thereof), including Cicero, *Orator* 74; Pliny, *Natural History* 35.73; and Quintilian 2.13.12–13.

20. Jeffrey Wills, *Repetition in Latin Poetry: Figures of Allusion* (Oxford: Clarendon Press, 1996), 3.

21. Shane Butler, *The Matter of the Page: Essays in Search of Ancient and Medieval Authors* (Madison: University of Wisconsin Press, 2011), 13–27.

22. Wills, *Repetition*, 357–61.

23. Ovid, *Metamorphoses* 11.44–47.

24. Virgil, *Aeneid* 7.759–60.

25. Theocritus, *Idylls* 1.71–75.

26. On refrain's tendency toward *non*meaning, see John Hollander, "Breaking Into Song: Some Notes on Refrain," in *Lyric Poetry: Beyond New Criticism*, ed. Chaviva Hošek and Patricia A. Parker (Ithaca, NY: Cornell University Press, 1985), 75.

27. Virgil, *Georgics* 4.523–27. A few editors insert quotation marks, though this undoes the ambiguity of the accusative, which is probably exclamatory in the middle instance, with the others offering something slightly more muted than direct speech: not so much "he cries 'Eurydice!'" as "he cries the name Eurydice."

28. Compare his doubts just before: "'tis not to me she speaks."

29. Similar but much more extensive commentary on the balcony scene concludes the appendix, "Dedicated to Derrida," of Adriana Cavarero, *For More Than One Voice: Toward a Philosophy of Vocal Expression*, trans. Paul A. Kottman (Palo Alto: Stanford University Press, 2005), 234–41. Cavarero is responding to Derrida's name-centered reading of the same in "Aphorism Countertime," in *Acts of Literature*, ed. Derek Attridge (London: Routledge, 1992), 416–33, where "he symptomatically misses the opportunity to deepen the theme of vocalic uniqueness along with Shakespeare" (236). Cavarero instead finds "a dialogue of voices" (241), though she concentrates on Romeo's voice rather than Juliet's.

30. On the sophistication of Virgil's better-known echoes in the *Eclogues*, see especially Phillip Damon, *Modes of Analogy in Ancient and Medieval Verse* (Berkeley: University of California Press, 1961), 281–90, and Brian W. Breed, *Pastoral Inscriptions: Reading and Writing Virgil's "Eclogues"* (London: Duckworth, 2006), 74–94.

31. Ovid, *Metamorphoses* 1.1–4.

32. Homer, *Iliad* 2.494–759. But regarding the playfully appropriate etymologies of Ovid's invented names, even closer Homeric parallels are to be found in the list of Nereids at *Iliad* 18.35–51 or that of Phaiakian nobles at *Odyssey* 8.111–19, on which see Elizabeth Minchin, "Lists and Catalogues in the Homeric Epics," in *Voice into Text: Orality and Literacy in Ancient Greece*, ed. Ian Worthington (Leiden: E. J. Brill, 1996), 10–11, who offers provocative suggestions about the special pressures (and listening pleasures) produced by Homeric lists in performance: "On the printed page lists and catalogues hold little attraction for most readers . . . But to hear a list song is, it seems, a very different experience" (5). A very different but complementary suggestion about superabundant enumeration in medieval descriptions, especially of markets, is given by Emma Dillon, *The Sense of Sound:*

Musical Meaning in France, 1260–1330 (New York: Oxford University Press, 2012), 73–75; following her lead, we can see Ovid's interrupted catalogue as a kind of overrunning of language by the multiplying (and ever louder!) dogs it seeks in vain to capture.

33. Ovid, *Metamorphoses* 3.206–25.

34. Cf. William S. Anderson, ed., *Ovid's "Metamorphoses": Books* 1–5 (Norman: University of Oklahoma Press, 1997), 360.

35. In speech, the first encounter with a proper name, generally underdetermined by the linguistic code has preceded it ("May I introduce Mr. *x*"), elicits maximum, syllable-by-syllable attention on the part of a listener; cf. Roman Jakobson and Morris Halle, *Fundamentals of Language*, 4th ed. (The Hague: Mouton, 1980), 3, opening their discussion of "phonology and phonetics."

36. Anderson, *Ovid's "Metamorphoses,"* 359. Cf. Mark Payne, *The Animal Part: Human and Other Animals in the Poetic Imagination* (University Of Chicago Press, 2010), 129: "an impressive metrical *tour de force*" that provides "one of the strangest moments in a poem full of strange and fantastic events." Discussion of Ovid's lists (though in relation to interpolation) in R. J. Tarrant, "The Soldier in the Garden and Other Intruders in Ovid's 'Metamorphoses,'" *Harvard Studies in Classical Philology* 100 (2000): 425–38.

37. Ovid, *Metamorphoses* 3.229–31.

38. Homer, *Odyssey* 17.291.

39. Quoted in P. Burmann, ed., *Publii Ovidii Nasonis Opera omnia* (Amsterdam: Apud R. et J. Westenios et G. Smith, 1727), 2:189. R. J. Tarrant, ed., *P. Ovidi Nasonis Metamorphoses* (Oxford: Oxford University Press, 2004), follows Heinsius in secluding the line; discussion of the general question of the poem's interpolations on pp. xxxiii–xxxiv; more in idem, "Toward a Typology of Interpolation in Latin Poetry," *Transactions of the American Philological Association* 117 (1987): 281–98; also "The Soldier," cited earlier. Missing the forest for the trees, Alessandro Barchiesi and Gianpiero Rosati, eds., Ovid, *Metamorfosi* (Milan: A. Mondadori, 2007), 2:162, insist, "alla luce dell'importanza del nome di Atteone nella storia," that the line is "da ritenere genuino."

40. Ovid, *Metamorphoses* 15.876–79.

41. On the passage and its "mouth" in the broader context of the Roman master trope of *fama*, see Philip Hardie, *Rumour and Renown: Representations of* Fama *in Western Literature* (Cambridge: Cambridge University Press, 2012), 167.

42. On the epilogue's construction of the poem as both cenotaph and tomb, see Philip Hardie, *Ovid's Poetics of Illusion* (Cambridge: Cambridge University Press, 2002), 91–97. In keeping with that book's broader themes, Hardie concludes that the reader's voice ultimately confirms the living author's absence from his "textual body."

43. Michel Foucault, "What Is an Author?" in *Language, Counter-Memory, Practice: Selected Essays and Interviews*, ed. Donald F. Bouchard, trans. Donald F. Bouchard and Sherry Simon (Ithaca, NY: Cornell University Press, 1977), 113–38.

44. Gertrude Stein, "Sacred Emily," in *Geography and Plays* (Boston: The Four Seas Company, 1922). Stein's first "Rose" is a woman's name.

45. Stewart Garrett, *Reading Voices: Literature and the Phonotext* (Berkeley: University of California Press, 1990), 26–27.

46. Shane Butler, "Beyond Narcissus," in *Synaesthesia and the Ancient Senses*, ed. Shane Butler and Alex Purves (Durham, UK: Acumen, 2013), 185–200.

47. For more, see Ezio Pellizer and Maurizio Bettini, *Il mito di Narciso: Immagini e racconti dalla Grecia a oggi* (Turin: G. Einaudi, 2003).

48. E.g., Shadi Bartsch, *The Mirror of the Self: Sexuality, Self-Knowledge, and the Gaze in the Early Roman Empire* (Chicago: University of Chicago Press, 2006), 89: "This whole episode is ripe for a Lacanian reading, given that the 'mirror stage' itself consists in a construction of identity based on a misprision (*méconnaissance*) of the image in the mirror." What is sure to remain the definitive Lacanian reading of the episode has since appeared in Micaela Janan, *Reflections in a Serpent's Eye: Thebes in Ovid's "Metamorphoses"* (Oxford: Oxford University Press, 2009), 114–55. On Lacan's mirror stage and Ovid, see Butler, "Beyond Narcissus."

49. Ovid, *Metamorphoses* 3.348.

50. Ovid himself advises a visit to the oracle and a reading of its famous inscription at *Ars amatoria* 2.497–500.

51. Adding to the fun is the erotic ambiguity of "knowing" (*noscere*), as observes Peter E. Knox, *Ovid's "Metamorphoses" and the Traditions of Augustan Poetry*, (Cambridge: Cambridge Philological Society, 1986), 20. For ways in which the episode also plays on the story of Oedipus (beginning with the blinding of Tiresias), see Ingo Gildenhard and Andrew Zissos, "Ovid's Narcissus (Met. 3.339–510): Echoes of Oedipus," *American Journal of Philology* 121.1 (2000): 129–47, with references to earlier discussions.

52. Homer, *Odyssey* 11.92–149; Sophocles, *Antigone* 988–1090; *Oedipus the King*

316–462; Euripides, *Bacchae* 170–369; *The Phoenician Women* 834–959.

53. The connection is explored by Maurizio Bettini, *Voci: Antropologia sonora del mondo antico* (Turin: G. Einaudi, 2008), 160–68.

54. Compare the prophecy made by a mysterious voice early in the Theban cycle, at 3.97–98 (*quid, Agenore nate, peremptum serpentum spectas? et tu spectabere serpens*), where, however, the sibilant alliteration serves clearly onomatopoeic ends.

55. Virgil, *Eclogues* 4.58–63. A connection to circulating collections of Sybilline oracles was first proposed by Roland G. Austin, "Virgil and the Sibyl," *Classical Quarterly* 21.2 (1927): 100–105.

56. Plato, *Timaeus* 71e–72a. Discussion in Giovanni Manetti, *Le teorie del segno nell'antichità classica* (Milan: Bompiani, 1987), 27–33. Compare Paul's instructions about the interpretation of *glossolalia* ("speaking in tongues") by ecstatic Christians in 1 Corinthians 12–14 (discussion in Bettini, *Voci*, 168–73), which may likewise lie at the root of modern imaginings of the Pythia at work.

57. Joseph Fontenrose, *The Delphic Oracle: Its Responses and Operations, with a Catalogue of Responses* (Berkeley: University of California Press, 1978), 196–228, delivers a full rebuttal of the older view; a recent summary, with more extensive comments on the tradition of Delphic verse responses, can be found in Hugh Bowden, *Classical Athens and the Delphic Oracle: Divination and Democracy* (Cambridge: Cambridge University Press, 2005), 12–39. Picking up on Fontenrose, Steven Connor, *Dumbstruck: A Cultural History of Ventriloquism* (Oxford: Oxford University Press, 2000), 47–74, places Delphic practice (including their erroneous reconstructions) at the heart of "the West's fascination with the nature and meaning of voice" (49). Among the topics in Connor's rich exploration (in a book that has become foundational reading in the field of voice studies) are the pointedly gendered ways in which the Pythia and her sanctuary embody the voice, as well as the ways in which she and other prophetesses dramatize both "the process of speech production" (56) and "the bursting apart of meaning" (66); he moves beyond Delphi itself to consider the Sybilline books (57–59) and to provide close readings of representations of prophetic Sibyls in Virgil and Lucan (59–69).

58. Georg Wilhelm Friedrich Hegel, *Lectures on the Philosophy of History*, trans. J. Sibree (London: Henry G. Bohn, 1857), 246–47.

59. Roland Barthes, *The Rustle of Language* (Berkeley: University of California Press, 1986), 79.

60. Calvert Watkins, *The American Heritage Dictionary of Indo-European Roots*, 3rd ed. (Boston: Houghton Mifflin Harcourt, 2011), 90. Varro, *Antiquitates Rerum Divinarum* fr. 107 (ed. Cardauns), explains that "the god Vaticanus, who presides over the beginnings of the human voice, gets his name from the fact that children, as soon as they are born, give out that sound that is in the first syllable of Vaticanus and so are described by the verb *vagire*, which expresses the sound of the voice just produced." On the application of the same verb to the sound made by hares, see Bettini, *Voci*, 81.

61. Ovid, *Metamorphoses* 3.380.

62. Ibid. 3.386–87.

63. Ibid. 3.382.

64. Ibid. 3.399.

65. Ibid. 3.425–26.

66. Ibid. 3.463. Ellen O'Gorman has made to me the brilliant observation that a further echo, between *ego* and *(im)ago*, suggests that Narcissus can distinguish between self and illusion at the level of word and concept, but that the distinction breaks down at the level of sound.

67. Summing up the scene of Narcissus beating his breast in despair, Frederick Ahl, *Metaformations: Soundplay and Wordplay in Ovid and Other Classical Poets* (Ithaca, NY: Cornell University Press, 1985), 239, remarks, "Soundplay runs riot." Here and throughout his study, however, Ahl largely limits himself to *meaningful* wordplay, such as palindromes and anagrams (especially apt, as he notes, in this scene of echoes and reflections) and words hidden within words to reinforce (or complicate) controlling images and themes.

68. Julia Kristeva, *Revolution in Poetic Language*, trans. Margaret Waller (New York: Columbia University Press, 1984), 79.

69. Ibid., 75, 79–80.

70. Ibid., 79, 86.

71. Ibid., 81.

72. Ibid., 29.

73. Roland Barthes, "Listening," in *The Responsibility of Forms: Critical Essays on Music, Art, and Representation*, trans. Richard Howard (New York: Hill and Wang, 1985), 245, 259, with the removal of the translator's added italics and the addition of the bracketed French, from Roland Barthes, "Écoute," in *L'obvie et l'obtus: Essais critiques III* (Paris: Éditions du Seuil, 1982), 229, which notes, 230, the essay's

original appearance in the *Encyclopédie Einaudi* (1976).

74. Carolyn Abbate, *Unsung Voices: Opera and Musical Narrative in the Nineteenth Century* (Princeton: Princeton University Press, 1991), ix, xv.

75. Ibid., 4.

76. Ibid., 6.

77. Ibid., 5.

78. Ovid, *Metamorfosi*, ed. Alessandro Barchiesi, 6 vols. (Milan: A. Mondadori, 2005–), of which the first five have appeared.

79. Alessandro Barchiesi, "Narrative Technique and Narratology in the Metamorphoses," in *The Cambridge Companion to Ovid*, ed. Philip R. Hardie (Cambridge: Cambridge University Press, 2002), 180.

80. E.g., Alessandro Barchiesi, "Voices and Narrative 'Instances' in the *Metamorphoses*," in *Speaking Volumes: Narrative and Intertext in Ovid and Other Latin Poets*, trans. Matt Fox and Simone Marchesi (London: Duckworth, 2001), 49–78; and "Music for Monsters: Ovid's Metamorphoses, Bucolic Evolution, and Bucolic Criticism," in *Brill's Companion to Greek and Latin Pastoral*, ed. Marco Fantuzzi and Theodore Papanghelis (Leiden: Brill, 2006), 403–26.

81. Barchiesi, *Speaking Volumes*, 8.

82. Gianpiero Rosati, "Narciso o l'illusione dissolta (Ovidio, *Metam.* III 339–510)," *Maia* 28 (1976), 90–92; reprised in *Narciso e Pigmalione: Illusione e spettacolo nelle Metamorfosi di Ovidio* (Florence: Sansoni, 1983), 30–34. Rosati promises not to treat Echo's story as "solo un momento secondario," but even his efforts at parity (the two figures "si contrappongono specularmente") stick to the optical paradigm (1976: 87; 1983: 26).

83. More subtly, Barbara Pavlock, *The Image of the Poet in Ovid's "Metamorphoses"* (Madison: University of Wisconsin Press, 2009), 14–37, begins her study with Narcissus but finds in him (and, suggestively, in Echo's lament) an image not of the poet of *Metamorphoses* but of his elegiac alter ego.

84. Ausonius 13.11 (ed. Green).

85. Joseph Loewenstein, *Responsive Readings: Versions of Echo in Pastoral, Epic, and the Jonsonian Masque* (New Haven: Yale University Press, 1984); discussion of the epigram, 25–26. Also, looking later, John Hollander, *The Figure of Echo: A Mode of Allusion in Milton and After* (Berkeley: University of California Press, 1981). The most extensive collection of Echo-inspired texts is that of Louise Vinge, *The Narcissus Theme in Western European Literature up to the Early 19th Century* (Lund:

Gleerups, 1967).

86. Boccaccio, *Genealogy of the Pagan Gods* 7.59.1. Note that an often-cited phrase from the relevant passage does not mean what it usually is said to mean; rather, Boccaccio reports that Narcissus "was loved by many nymphs, but especially by Echo, a nymph of Mt. Parnassus" (*a nynphis pluribus amatus est, sed ab Echo Parnasi nynpha potissime*).

87. Enterline, *Rhetoric of the Body*, 56, noting that this is itself constructed of the repeated words and figures of other poets, a matter to which we will turn.

88. Victoria Rimell, *Ovid's Lovers: Desire, Difference and the Poetic Imagination* (Cambridge: Cambridge University Press, 2006), 40. Enterline, *Rhetoric of the Body*, 79–83, likewise finds a figure for poetry in Medusa, though in terms of her silent and silencing *os* (both "mouth" and "face").

89. Rimell, *Ovid's Lovers*, 1.

90. Rosati, *Narciso e Pigmalione*, 22–23.

91. Enterline, *Rhetoric of the Body,* 87, invoking the *Heroides*; for his resemblance to Apollo in the *Ars amatoria*, see Pavlock, *Image of the Poet*, 21–24. More generally, John Brenkman, "Narcissus in the Text," *Georgia Review* 30 (1976): 325, observes, "Tiresias's enigmatic forecast, which 'comes true' in the course of the narrative, puts him in a special relation to its truth; the narrative can be said to elaborate or unfold the truth of Tiresias's prophetic statement.... And to the extent that the relation between his language and truth or meaning is analogous to the relation posed between the narrator and the narrative, Tiresias is even a 'figure of the poet.'"

92. Homer, *Odyssey* 12.189–91.

93. Indeed, as Pietro Pucci, "The Song of the Sirens," in *The Song of the Sirens: Essays on Homer* (Lanham, MD: Rowman & Littlefield, 1998): 1–9, reveals, even their "diction 'reproduces'—so to speak—the diction of the *Iliad*" (1).

94. Hardie notes the Ennian intext (and suggests a further reference to Anna's address to Dido at Virgil, *Aeneid* 4.685) in Hardie, *Ovid's Poetics of Illusion*, 94-5. Echo haunts that book's chapter on "Narcissus: The Mirror of the Text" (143–72); tellingly, however, she does so as the character rather than the sound, producing not just words but meaning ("the repetitions in this story *do* convey her intentions," 165) and so becoming a figure for language itself: "In Lacanian terms Narcissus is inflicted with the delusion of a true presence in the realm of the Imaginary as punishment for his refusal to enter a relationship with Echo in the

realm of the Symbolic, conducted through the mediations of language" (164–65).

95. Ennius, frag. 46 (ed. Courtney), quoted by Cicero, *Tusculan Disputations* 1.34; Virgil, *Georgics* 3.9.

96. E.g., by Gian Biagio Conte, *Latin Literature: A History*, trans. Joseph B. Solodow (Baltimore: Johns Hopkins University Press, 1994), 82: Ennius's "alliterating style, typical of the earliest *carmina*, was native to Latin. We find it in the proverbs, laws, and sacral formulas, and it passes into Naevius's Saturnians and into the verses of comedies by Naevius, Plautus, or Caecilius. Ennius imported it into the hexameter, imposing thus upon a Greek verse the effects of a specifically Roman style (in Greek poetry alliteration plays no appreciable role)." Conte interestingly supposes that alliteration gave the freer meters of Latin poetry before Ennius "a kind of regularity, a kind of rhythmic armature"; he suggests that Ennian alliteration is therefore overkill ("monotonous and heavily cadenced") and that a later poet like Virgil often will alliterate "to sound Ennian, to give his verses the ring of a poetry more ancient and traditional, now far away." But this can hardly account for all Virgilian alliteration, far less for its importance to Ovid. It would be better to say that Ennian and other early associations helped to emphasize alliteration's elemental (rather than simply archaizing) ring.

97. Ovid does not explicitly mention this oracular sequel but probably alludes to it (as a cave?) by having Apollo turn the head of Orpheus, its mouth gaping wide, to stone (*congelat et patulos, ut erant, indurat hiatus*, 11.60). The oracle is described by Lucian, *The Ignorant Book Collector* 11–12 and the head itself is depicted on several Greek vases: see Walter Burkert, *Homo Necans: The Anthropology of Ancient Greek Sacrificial Ritual and Myth* (Berkeley: University of California Press, 1983), 202; image of the most famous of these in C. D. Bicknell, "Some Vases in the Lewis Collection," *Journal of Hellenic Studies* 41 (1921), plate 12. "Oracular edifices are frequently said to have a στόμα or στόμιον," notes Tim Whitmarsh, "Greek and Roman in Dialogue: The Pseudo-Lucianic Nero," *Journal of Hellenic Studies* 119 (1999), 154 n. 82, with references.

98. Susan Sontag, "Against Interpretation," in *Against Interpretation and Other Essays* (New York: Dell, 1967), 14.

TRACK THREE: EINE KLEINE NACHTMUSIK

1. Gellius, *Attic Nights* 19.9.1.

2. Ibid. 19.9.2. Gellius usually party-crashed in the company of Favorinus;

see William A Johnson, *Readers and Reading Culture in the High Roman Empire: A Study of Elite Communities* (New York: Oxford University Press, 2010), 103. For a complete introduction to Gellius and friends, see Leofranc Holford-Strevens, *Aulus Gellius: An Antonine Scholar and His Achievement*, rev. ed. (Oxford: Oxford University Press, 2003).

3. Gellius, *Attic Nights* 19.9.3.

4. Ibid.

5. Ibid. 19.9.4.

6. Ibid. 19.9.5.

7. Marcel Proust, *Swann's Way*, trans. C.K. Scott Moncrieff, Terence Kilmartin, and D.J. Enright (New York: Modern Library, 1992), 63.

8. The statue (real or invented) is described in an epigram by Leonidas of Tarentum preserved in *The Greek Anthology* 16.306.

9. Gellius, *Attic Nights* 19.9.3.

10. The term may also suggest that they understood what they were singing. See 16.6; 18.5; Johnson, *Readers*, 129.

11. Gellius, *Attic Nights* 19.9.6. I give here (and translate on the basis of) the Greek text of West, minus his supplements and restored adscripts.

12. Ibid. 19.9.7.

13. Ibid. 19.9.8.

14. Ibid. 19.9.9.

15. Ibid. 19.9.10–14.

16. Michèle Lowrie, *Writing, Performance, and Authority in Augustan Rome* (Oxford: Oxford University Press, 2009), uses this Virgilian beginning to launch her exploration of the relationship between "song" and writing in Latin literature's negotiation of "presence" and so of more pointedly ideological values, like the "authority" of her title: "For literary texts in Augustan Rome, writing and song, presence and absence are thematically inextricable. I start by identifying song with presence and writing with absence as a heuristic gesture on the way towards differentiating aspects of a larger practice whose terms easily shift into one another depending on the particular point being made" (7). This "aesthetics of presence," however, leaves strangely little room for ordinary aesthetic quantities like literary pleasure, and the book's lack of a convincing general theory of artistic materiality itself (ancient or otherwise) tends to lead Lowrie to overstate both the Augustan (or Roman) specificity and ideological thrust of her sources (e.g., "in the Augustan

age there is a shared yearning in artistic and political representation for presence," 23). For a helpful review, see Luke Roman, *Classical Review* 61.1 (2011): 118–21.

17. Gellius, *Attic Nights* 19.9.10.

18. Ibid. 19.9.5.

19. Throughout the *Attic Nights*, Gellius often uses *vox* in the former sense interchangeably with *verbum*, *vocabulum*, etc.—sometimes in close proximity to the use of *vox* for "voice," as at 5.7, where *vox* means one thing in the title but then another as the chapter begins. In a few passages (e.g., 2.3.1), however, *vox* has a slightly narrower sense, designating a word's "pronunciation" (cf. *OLD* uox 6, "the auditory effect of a word, etc., sound," giving the somewhat complicated example of 16.18.4) or, perhaps better, "material form" (since it seems to refer as much to spelling as to pronunciation).

20. Dolly Parton, *Halos & Horns* (Nashville: Sugar Hill, 2002).

21. The most famous version of this teleology, especially regarding its instrumental end point, is probably that of Eduard Hanslick in his *Vom Musikalisch-Schönen: Ein Beitrag zur Revision der Ästhetik der Tonkunst*, first published in 1854, translated as Eduard Hanslick, *The Beautiful in Music: A Contribution to the Revisal of Musical Aesthetics*, trans. Gustav Cohen (London: Novello and Co., 1891): "The fundamental difference [between speech and music] consists in this: while *sound* in *speech* is but a sign, that is, a *means* for the purpose of expressing something which is quite distinct from its medium; *sound* in *music* is the *end*, that is, the ultimate and absolute object in view" (94). See further Carl Dahlhaus, *The Idea of Absolute Music* (Chicago: University of Chicago Press, 1989).

22. For example, instrumental music on the *kithara* or *aulos*, without singing, comprised two of the five performative competitions at the Panathenaia; discussion in Gregory Nagy, "Language and Meter," in *A Companion to the Ancient Greek Language*, ed. Egbert J. Bakker (Chichester, UK: Wiley-Blackwell, 2010), 376.

23. The *Oxford Latin Dictionary* actually begins here, deferring "song" to the word's second meaning. Thomas Habinek, *The World of Roman Song: From Ritualized Speech to Social Order* (Baltimore: Johns Hopkins University Press, 2005), as his title suggests, follows this line to far-reaching conclusions about Roman culture; see also, more briefly, his "Singing, Speaking, Making, Writing: Classical Alternatives to Literature and Literary Studies," *Stanford Humanities Review* 6.1 (1998): 65–75.

24. Beyond cognates like *cantus* (which admits similar ambiguities and properly

means "singing," that is, song as a category rather than an item), metonyms like *modi*, and the rare Greek import *melos*.

25. Gregory Nagy, *Pindar's Homer: The Lyric Possession of an Epic Poet* (Baltimore: Johns Hopkins University Press, 1990), 17–43.

26. This standard, simplified description sidesteps controversies both ancient and modern; for something more careful, see W. Sidney Allen, *Accent and Rhythm: Prosodic Features of Latin and Greek: A Study in Theory and Reconstruction* (Cambridge: Cambridge University Press, 1973), 46–73, 255–59.

27. M.L. West, ed., *Carmina Anacreontea* (Leipzig: Teubner, 1984), xiv.

28. A.M. Dale, *Collected Papers* (Cambridge: Cambridge University Press, 1969), 161; M.L. West, *Ancient Greek Music* (Oxford: Clarendon Press of Oxford University Press, 1992), 130. But there could be divergence: see Luigi Battezzato, "Metre and Music," in *The Cambridge Companion to Greek Lyric*, ed. Felix Budelmann (Cambridge: Cambridge University Press, 2009), 143.

29. I borrow the example from James Anderson Winn, *Unsuspected Eloquence: A History of the Relations Between Poetry and Music* (New Haven: Yale University Press, 1981), 3.

30. Gertraud Frenk-Oczlon and August Fenk, "Musical Pitch in Nonsense Syllables: Correlations with the Vowel System and Evolutionary Perspectives," in *Proceedings of the Seventh Triennial Conference of the European Society for the Cognitive Sciences of Music*, ed. J. Louhivou et al. (Jyväskylä, Finland: University of Jyväskylä, 2009), 110–13. Cf. George W. Boswell, "Pitch: Musical and Verbal in Folksong," *Yearbook of the International Folk Music Council* 9 (1977): 80–88. Maurizio Bettini, *Voci: Antropologia sonora del mondo antico* (Turin: G. Einaudi, 2008), 96–100, interestingly suggests that some of the features of yodeling can perhaps be found in the rustic practices captured by the Latin verb *iubilare*.

31. Herbert Weir Smyth, *Greek Grammar* (Cambridge, MA: Harvard University Press, 1956), 37 (§152).

32. Gregory Nagy, "Language and Meter," 383–84, borrowing "melodic accentuation" from W. Sidney Allen, *Vox Graeca: A Guide to the Pronunciation of Classical Greek*, 3rd ed. (Cambridge: Cambridge University Press, 1987), 116–31. For cautions, drawn from comparisons with other traditions, about the consistency of pitch accent in speech and its observation in song, see Warren D. Anderson, *Music and Musicians in Ancient Greece* (Ithaca, NY: Cornell University Press, 1994), 45–47.

33. A potential additional element regards the visible and audible marking of

rhythm by the (dancing) body, theorized in antiquity as *arsis* ("raising") and *thesis* ("lowering"), which can be experienced as a kind of "stress." However, the relationship of these movements to the vocal sounds of song or speech is extremely controversial; see Allen, *Accent and Rhythm*, 100, 275–79. A more fruitful approach to the question in this particular case would be to consider how, for the right reader, the verses quoted by Gellius may also partially notate the dancing that is an explicit part of their context here (and from which our focus on "music" has somewhat unnaturally disembodied them).

34. See, most recently, Stefan Hagel, *Ancient Greek Music: A New Technical History* (Cambridge: Cambridge University Press, 2010).

35. Battezzato, "Metre and Music," 143, describes opposing directions taken by the evidence: "We know that there was only a partial correspondence between word accent and melodic pattern and we see that accent patterns changed from strophe to strophe, even if the same melody was presumably repeated in each strophe. Some astrophic Hellenistic poems for which musical notation has survived display a high degree of coherence between word accent and melodic pattern."

36. For an attempt to trace the evolution of the lyrics, see Theodor Bergk, ed., *Poetae Lyrici Graeci*, 4th ed. (Leipzig: B. G. Teubner, 1882), 3:298–300.

37. West, *Ancient Greek Music*, 178.

38. *Moduli* refers to the prose of Plato in the fascinating discussion of translating him into Latin in Gellius, *Attic Nights* 17.20.

39. Wolfgang Amadeus Mozart, *Don Giovanni*, the New Philharmonia Orchestra and Chorus, Angel Records SDL 3700 (1–4).

40. On the complex efforts, instead, of performance artist Juliana Snapper to *unlearn* her operatic training and otherwise disrupt its effects in order to restore priority to "the physicality of her instrument," see Nina Sun Eidsheim, "Sensing Voice: Materiality and the Lived Body in Singing and Listening," *Senses & Society* 6.2 (2011), 136–38.

41. Cf. Gregory Nagy, "Did Sappho and Alcaeus Ever Meet? Symmetries of Myth and Ritual in Performing the Songs of Ancient Lesbos," in *Literatur und Religion I: Wege zu einer mythisch-rituellen Poetik bei den Griechen*, ed. A. Bierl, R. Lämmle, and K. Wesselmann (Berlin: Walter de Gruyter, 2007), 211–70.

42. On the classical meaning of *suavis* (and *dulcis*, usually preferred for "sweet" in the narrow sense of "sweet-tasting"), see A. H. Mamoojee, "'Suavis' and 'Dulcis': A Study of Ciceronian Usage," *Phoenix* 35.3 (1981): 220–36, who notes, "In

describing the orator's voice, *suavis* is consequently much preferred to *dulcis* (10 to 1), as being more expressive in distinguishing the pleasant or modulated aspect of the voice from its other aspects, such as its being loud (*magna*), clear (*clara*, *splendida*), sonorous (*canora*), or full (*plena*)" (232).

43. Compare the superlative voice of an actor noted by Gellius at *Attic Nights* 6.5.1: *Histrio in terra Graecia fuit fama celebri, qui gestus et vocis claritudine et venustate ceteris antestabat.* At 16.19.13, Arion sings *voce sublatissima*, for which the *OLD* suggests very "high-pitched." A different "role" of the voice appears in 3.5.2, where a *vox infracta* is typecast as an unmistakable sign of a *cinaedus*.

44. See Chapter 2, n. 19.

45. Douglas E. Gerber, *A Companion to the Greek Lyric Poets* (Leiden: Brill, 1997), 198–201; Felix Budelmann, "Anacreon and the Anacreontea," in *The Cambridge Companion to Greek Lyric*, ed. Felix Budelmann (Cambridge: Cambridge University Press, 2009), 227–39; Patricia A. Rosenmeyer, *The Poetics of Imitation: Anacreon and the Anacreontic Tradition* (Cambridge: Cambridge University Press, 1992), 12–36; Florence Dupont, *The Invention of Literature: From Greek Intoxication to the Latin Book*, trans. Janet Lloyd (Baltimore: Johns Hopkins University Press, 1999), 51–100, with discussion of the poem quoted by Gellius, 72–75. On the statue, see Flemming Johansen, *Greek Portraits: Ny Carlsberg Glyptotek* (Copenhagen: Ny Carlsberg Glyptotek, 1992), 18–21, and Alan Shapiro, *Re-Fashioning Anakreon in Classical Athens* (Berlin: Wilhelm Fink, 2012), both with additional bibliography. Shapiro reads the infibulation as a mark of sincere restraint and the statue as a depiction of Anacreon as "a model of the noble *erastes*" (46), but he conveniently reviews (22) other interpretations. On infibulation for the sake of the voice, see Edith Hall, "The Singing Actors of Antiquity," in *Greek and Roman Actors: Aspects of an Ancient Profession*, ed. Pat Easterling and Edith Hall (Cambridge: Cambridge University Press, 2002), 23–24.

46. Roland Barthes, "The Grain of the Voice," in *Image, Music, Text*, trans. Stephen Heath (New York: Hill and Wang, 1977), 179–89.

47. See the previous chapter, and Julia Kristeva, *Revolution in Poetic Language*, trans. Margaret Waller (New York: Columbia University Press, 1984), 86–89.

48. Barthes, "The Grain of the Voice," 182.

49. Donatus, *Ars grammatica major* (Keil, ed., *Grammatici Latini*, 4:367). For an accessible discussion of this and the following texts, see Blair Sullivan, "The Unwritable Sound of Music: The Origins and Implications of Isidore's Memorial

Metaphor," in *The Echo of Music: Essays in Honor of Marie Louise Göllner*, ed. Blair Sullivan (Warren, MI: Harmonie Park Press, 2004), 39–53; a shorter summary in Martin Irvine, *The Making of Textual Culture: 'Grammatica' and Literary Theory, 350–1100* (Cambridge: Cambridge University Press, 2006), 91–97. On their impact on the theory and practice of music into the Middle Ages, see Emma Dillon, *The Sense of Sound: Musical Meaning in France, 1260–1330* (New York: Oxford University Press, 2012), 36–43. For more on their ancient theoretical context, see Marcia L. Colish, *The Stoic Tradition from Antiquity to the Early Middle Ages* (Leiden: Brill, 1985), 321–27; D.M. Schenkeveld, "Studies in the History of Ancient Linguistics IV. Developments in the Study of Ancient Linguistics," *Mnemosyne* 43.3–4 (1990): 289–306; Valahfridus [Wilfried] Stroh, "De uocis definitione quadam Stoica," in *Mousopolos Stephanos: Festschrift für Herwig Görgemanns*, ed. Manuel Baumbach, Helga Köhler, and Adolf Martin Ritter (Heidelberg: C. Winter, 1998), 443–52; W. Ax, "Zum de voce-Kapitel der römischen Grammatik: Eine Antwort auf Dirk M. Schenkeveld und Wilfried Stroh," in *Grammatical Theory and Philosophy of Language in Antiquity*, ed. Pierre Swiggers and Alfons Wouters (Leuven: Peeters Publishers, 2002), 19–121. (The last three offer discussion of the meaning of *quantum in ipso est*.) Further dissection of the basic typology and of still finer distinctions is provided by Jesús Luque Moreno, "Voces. La clasificación de los sonidos en el mundo antiguo: I. Los gramáticos," *Voces* 7 (1996): 9–44.

50. Probus, *Instituta artium* (Keil, *Grammatici Latini*, 4:47).

51. The corresponding Greek terms are *phōnē enarthros* and *phōnē* (or *psophos*) *anarthros*.

52. Diomedes, *Ars Grammatica* 2 (Keil, *Grammatici Latini*, 1:420).

53. Varro, *On the Latin Language* 7.103-4.

54. The relation of such words (as "natural signs") to language in general receives its most famous ancient discussion in Plato's *Cratylus*, the legacy of which is the subject of Gérard Genette, *Mimologics*, trans. Thaïs E. Morgan (Lincoln: University of Nebraska Press, 1995).

55. This and the following survey of sound-words is synthesized from the grammatical texts already cited, especially Probus, plus Marius Victorinus, *Ars grammatica* 1 (Keil, *Grammatici Latini*, 6:4–5).

56. For more on ancient catalogues of animal sounds, see Bettini, *Voci*, 9–35.

57. Aristotle, *On Interpretation* 1.4.16b; 2.16a. See Chapter 1.

58. On which see, for starters, Tobias Reinhardt, "Epicurus and Lucretius on

the Origins of Language," *Classical Quarterly* 58.1 (2008): 127–40. I offer some observations on the Lucretian side of things in *The Matter of the Page: Essays in Search of Ancient and Medieval Authors* (Madison: University of Wisconsin Press, 2011), 51–54.

59. Gellius, *Attic Nights* 5.15.

60. Homer, *Iliad* 18.368–617.

61. Ibid. 18.571.

62. A similar connection in Virgil, between the description of the Orpheus cup in *Eclogue* 3 and the description the voice of Silenus in *Eclogue* 6, is delicately explored by Brian W. Breed, *Pastoral Inscriptions: Reading and Writing Virgil's Eclogues* (London: Duckworth, 2006), who suggests that both offer "the creation of voice as a phenomenal presence within the frame of representation" (77). Breed tends to insist that "phenomenal presence" is primarily (a marker of) absence, but his careful analysis of the "ecphrastic voice" (78–81) amounts to a case that both visual and vocal ecphrasis are even more compelling as symptoms of poetry's material presence.

63. Pantelis Michelakis, "Greek Lyrics from the Renaissance to the Eighteenth Century," in *The Cambridge Companion to Greek Lyric*, ed. Felix Budelmann (Cambridge: Cambridge University Press, 2009), 342–46.

64. Simon Hornblower and Antony Spawforth, eds., *The Oxford Classical Dictionary*, 4th ed. (Oxford: Oxford University Press, 2012), 77.

65. *Anacreontea* 32.

66. Pierre de Ronsard, *Oeuvres complètes*, ed. Jean Céard, Daniel Ménager, and Michel Simonin (Paris: Gallimard, 1993), 1:705–6. Rosenmeyer, *Poetics of Imitation*, 231, calls attention to these changes but supposed that Ronsard thereby "modernizes ancient customs and makes them relevant to contemporary French society."

67. James I. Porter, *The Origins of Aesthetic Thought in Ancient Greece: Matter, Sensation, and Experience* (Cambridge: Cambridge University Press, 2010), 454; Jesper Svenbro, *La parole et le marbre: Aux origines de la poétique grecque* (Lund: Lund University, 1976). Porter ends his own transformative study with the sublimely material (and materially sublime) voice, culminating (523) in a poetic meditation on the fleshy quanta of human expression.

TRACK FOUR: ARE YOU EXPERIENCED?

1. Friedrich Nietzsche, *The Birth of Tragedy and Other Writings*, ed. Raymond Geuss and Ronald Speirs, trans. Ronald Speirs (Cambridge: Cambridge University Press, 1999), 109 (Chapter 23).

2. Ibid., 108 (Chapter 23).

3. William Huck, "Tosca: Four Callas Toscas," *Opera Quarterly* 2.3 (1984): 178.

4. Ibid.

5. On the general controversy about the respective merits of the two singers, especially in relation to their Violettas, see Carolyn Abbate, *In Search of Opera* (Princeton: Princeton University Press, 2001), 45.

6. As she does in a final radio interview, available at http://youtu.be/g1DomEpyoCY.

7. W. J. Henderson, "Performance of Puccini's 'Tosca' at the Metropolitan Opera House," *New York Times*, February 10, 1901.

8. Friedrich A. Kittler, *Gramophone, Film, Typewriter*, trans. Geoffrey Winthrop-Young and Michael Wutz (Stanford: Stanford University Press, 1999).

9. The extremely recent (and emotionally close to home) setting of *The Capture of Miletus* by Phrynichus won a fine and a ban on future performances (Herodotus, *History* 6.21); so, too, was the relatively recent (but still past) setting of *The Persians* of Aeschylus exceptional.

10. Abbate, *In Search of Opera*, 6.

11. On this last point, I am adapting an argument about opera by Michal Grover-Friedlander, *Vocal Apparitions: The Attraction of Cinema to Opera* (Princeton: Princeton University Press, 2005), 3–8, extended in *Operatic Afterlives* (New York: Zone Books, 2011), where she notes, "Endless dying is what opens to the dimension of immortality. It forms a kinship among works that subverts the recurring absoluteness of their individual deaths. Each opera kills anew, thereby exposing the failure of death to master song" (17).

12. Theodor W. Adorno, "Opera and the Long Playing Record," in *Essays on Music*, ed. Richard D. Leppert, trans. Susan H. Gillespie (Berkeley: University of California Press, 2002), 285. The same volume includes the earlier essays "The Curves of the Needle" and "The Form of the Phonographic Record." The three pieces are brief, and their thought is sometimes frustratingly incomplete and elusive, but they are rich in provocations, such as this from "The Form": "For it is not in the play of the gramophone as a surrogate for music but rather in the

phonograph record as a thing that its potential significance—and also its aesthetic significance—resides" (278).

13. Aristotle, *Poetics* 13.1452b–1453a.

14. Most influentially in Jean-Pierre Vernant and Pierre Vidal-Naquet, *Myth and Tragedy in Ancient Greece* (New York and London: Zone, 1990), first published in French in two volumes in 1972 and 1986. Their thesis explicitly underpins, for instance, Simon Goldhill, "The Language of Tragedy: Rhetoric and Communication," in *The Cambridge Companion to Greek Tragedy*, ed. P. E. Easterling (Cambridge: Cambridge University Press, 1997), 127–50, who further explores the sources of tragedy's "particular register of language."

15. Mary Francis Gyles, "Nero Fiddled While Rome Burned," *Classical Journal* 42.4 (1947): 211–17. The same author attempts a sympathetic portrait of Nero the musical artist in "Nero: Qualis Artifex?" *Classical Journal* 57.5 (1962): 193–200.

16. Suetonius, *Life of Nero* 20. The same passage famously describes the physical regimen adopted by Nero for the sake of his voice, including the use of chest weights, enemas and purgatives, and abstinence from certain foods; cf. Pliny, *Natural History* 34.166. On Nero's practices and the broader context of ancient training and treatment of the voice, see Andrew Barker, "*Phōnaskia* for Singers and Orators: The Care and Training of the Voice in the Roman Empire," in *La musica nell'Impero romano: Testimonianze teoriche e scoperte archeologiche*, ed. Eleonora Rocconi (Pavia: Pavia University Press, 2010), 11–20. Timothy Power, *The Culture of Kitharōidia* (Washington, DC: Center for Hellenic Studies, 2010), 1–181, is now the definitive study of Nero's solo singing, which the author exploits as a way into the history of the art form itself, which he lavishly reconstructs (though he offers only brief commentary on the reported qualities of Nero's voice, 7–8, 121). More generally on Nero the performer, see Edward Champlin, *Nero* (Cambridge, MA: Harvard University Press, 2003), 53–83.

17. Virgil, *Eclogues* 10.37–38.

18. Suetonius, *Life of Nero* 51.

19. Ibid. 49.

20. More generally, Suetonius inherits and reflects a preoccupation with imperial voices pitilessly exploited, regarding the voice of the deceased Claudius, by Seneca in the *Apocolocyntosis*, a matter extensively explored by Josiah Osgood, "The *Vox* and *Verba* of an Emperor: Claudius, Seneca and *Le Prince Ideal*," *Classical Journal* 102.4 (2007): 329–53. As Osgood notes (345–46), the same satire has

Apollo praise Nero as his "lesser in neither song nor voice" (4.1).

21. Pseudo-Lucian, *Nero* 6. Conveniently available in M.D. Macleod, ed., *Lucian, with an English Translation* (London: Heinemann, 1967), 8:508–21, with discussion of its date and attribution (Macleod favors a Philostratus of the third century CE), 505–7, on which see, in agreement, Tim Whitmarsh, "Greek and Roman in Dialogue: The Pseudo-Lucianic Nero," *Journal of Hellenic Studies* 119 (1999): 142–60; both provide references to other discussions. Whitmarsh reads this particular passage in the context of the dialogue's broader themes of the voice (151–58). *Barus* here probably means "low," but for the word's range of meanings, see Maarit Kaimio, *Characterization of Sound in Early Greek Literature* (Helsinki: Societas Scientiarum Fennica, 1977), index, s. v. βαρύς. This and other vocal descriptors can also be compared in the convenient chart of Guy Lachenaud, *Les routes de la voix: L'antiquité grecque et le mystère de la voix* (Paris: Les Belles Lettres, 2013), 105–11.

22. R.M. Frazer, "Nero the Singing Animal," *Arethusa* 4.2 (1971): 215–18, on Plutarch, *Moralia* 567e–568a.

23. Cicero, *On the Nature of the Gods* 2.146.

24. M.L. West, *Ancient Greek Music* (Oxford: Oxford University Press, 1992), 42–43, similarly contrasts Nero's *vox exigua et fusca* (he translates the adjectives as "slight and husky") to "the commendatory adjective most regularly applied to the singing voice in Greek," namely, *ligus* or *liguros*, usually translated, in its positive sense, as "clear." West quotes the ambitious attempt to define this sound by Aristotle (attributed), *On Things Heard* 804a (where it is that of grasshoppers and nightingales, *inter alia*) but concludes, "We had better not try to extract too much from this.... The essential quality expressed by *ligys* seems to be clarity and purity of sound, free from roughness or huskiness." On the use of *ligus* to describe seemingly varied voices in Homer, see Silvia Montiglio, *Silence in the Land of Logos* (Princeton: Princeton University Press, 2000), 74–77. "By contrast," Montiglio notes, "in the classical age *ligus* is applied specifically to kinds of music and utterances in which sounds prevail over words" (148), while speech is called instead *lampros* ("luminous"), which "applies to the *logos* as well as to the speaker's voice" (149). See also Kaimio, *Characterization of Sound*, index, s. v. λιγυρός, λιγύς.

25. Apuleius, *De mundo* 33.

26. Aeschylus, *Seven Against Thebes* 859–60.

27. Pliny the Elder, *Natural History* 28.58.

28. Aristotle (attributed), *Problems* 30.1.953a (trans. E.S. Forster). On this passage's relationship to discussions of melancholy that are securely Aristotle's, see Philip J. van der Eijk, *Medicine and Philosophy in Classical Antiquity: Doctors and Philosophers on Nature, Soul, Health and Disease* (Cambridge: Cambridge University Press, 2005), 139–68.

29. Dio Cassius, *Roman History* 61.20.2, vol. 8, p. 78 (Loeb). The (neuter) adjectives modify *phōnēma*, "vocal sound."

30. On the relationship between the sources, see Henry Bardon, *Les empereurs et les lettres latines d'Auguste à Hadrien*, 2nd ed., rev. (Paris: Les Belles Lettres, 1968), 197; K.R. Bradley, *Suetonius' Life of Nero: An Historical Commentary* (Brussels: Latomus, 1978), 123.

31. William T. Stearn, *Botanical Latin: History, Grammar, Syntax, Terminology and Vocabulary*, 4th ed. (Newton Abbot, UK: David & Charles, 1992), 244.

32. Plato, *Republic* 9.584e–585a.

33. *Thesaurus Linguae Latinae*, "fuscus" 6.1.1654.49.

34. Indeed, the first example listed by Liddell and Scott (Aristotle, *Topics* 1.15.106b7) actually uses the color word *phaios*, somewhere between "white" (*leukos*) and "black" (*melas*), to illustrate the *lack* of any analogously intermediate term for sound, except perhaps for *somphos* ("spongy," "muffled"). The definition seems instead to come from the second example, Aristotle (attributed), *On Things Heard* 802a1, where the discussion is of clearness and the subsequent explanation is intriguing: "Such, then, are the conditions of clearness in the voice. So voices which are called 'grey' are generally considered no worse than those which are called 'white.' For voices which are rather harsh (*trachuterai*) and slightly confused and have not any very marked clearness are the fitting accompaniment of outbreaks of passion and of advancing years, and at the same time, owing to their intensity, they are less under control; for what is produced by violent exertion is not easily regulated, for it is difficult to increase or decrease the strength of the sound at will" (trans. T. Loveday and E.S. Forster; for a rather different rendering of the key terms, see Barker, *Greek Musical Writings*, 2:103).

35. William Bedell Stanford, *Greek Metaphor: Studies in Theory and Practice* (Oxford: B. Blackwell, 1936), 47–62. Demetrius, *On Style* 2.86. On the general phenomenon of lightness and darkness as stylistic metaphors and judgments, see Larue Van Hook, *The Metaphorical Terminology of Greek Rhetoric and Literary Criticism* (Chicago: University of Chicago Press, 1905), 14–15.

36. Stanford, *Greek Metaphor*, 59. Pauline A. LeVen, "The Colors of Sound: *Poikilia* and Its Aesthetic Contexts," *Greek and Roman Musical Studies* 1 (2013), quoting Stanford with (qualified) approval (232, n. 12), offers an engaging exploration of the application of the Greek adjective *poikilos* ("variegated" is one basic translation) to sound. "Rather than being transferred from one realm to another, the adjective shows the continuity between the senses regardless of their object" (238–39); as LeVen reveals, part of what makes *poikilos* an especially interesting example of the crossing of sensory realms is the fact that, already within a single one, it designates a multiplicity of effects.

37. Quintilian, *The Orator's Education* 11.3.1.

38. Ibid. 11.3.15.

39. Ibid. 11.3.171.

40. Cicero, *Brutus* 141.

41. Seneca the Elder, *Controversiae* 1.Pr.16.

42. Cicero, *Brutus* 141–42.

43. Ovid, *Metamorphoses* 11.52–53.

44. Shadi Bartsch, *Actors in the Audience: Theatricality and Doublespeak from Nero to Hadrian* (Cambridge, MA: Harvard University Press, 1994), 224–25 n. 6, provides a convenient list; she also briefly discusses (224 n. 4) the nature of these performances, noting that "the brief notices in the sources do not always permit us to distinguish Nero's appearances as *citharoedus*, performer of tragic arias to the lyre, from those as *tragoedus*, or tragic actor; nor is it entirely clear to what degree the latter approximated bona fide tragic productions." Much more in Power, *Culture of Kitharōidia*, who observes that there may not have been much difference in Nero's repertoire, "in its performative techniques and modes of self-presentation," between the lyric songs (*nomoi*) of *kitharōidia* and the "tragic acting plus solo singing" of *tragōidia* (143). An enigmatic reference in a fragment of Lucilius (frag. 594, ed. Warmington; frag. 567, ed. Marx) to a tragic actor who "ruins the music" (*carmina perdit*) by making use of an Orestes who is *rausurus* (a *hapax legomenon* that presumably means "becoming hoarse") perhaps suggests the opposite, i.e., a long-standing interest in finding an appropriately "dark" voice for what would become one of Nero's celebrated roles.

45. Suetonius, *Life of Nero* 21.3.

46. Seneca, *Hercules (Furens)* 1138–48.

47. For the suggestion that the performance was "based on the *Herakles* of

Euripides," see Margarethe Billerbeck, "Hercules Bound: A Note on Suetonius, *Nero* 21.3," *American Journal of Philology* 102.1 (1981): 54–57.

48. Favoring the hypothesis of musical accompaniment, however, is Edith Hall, "The Singing Actors of Antiquity," in *Greek and Roman Actors: Aspects of an Ancient Profession*, ed. Pat Easterling and Edith Hall (Cambridge: Cambridge University Press, 2002), 26. The controversy around the performability of Senecan tragedy would be long to review; for a sensible compromise (with bibliography from both sides), insisting that the plays belong, "if anything does, to the category of Roman performance," but leaving open some questions about their performative particulars, see A. J. Boyle, *Tragic Seneca: An Essay in the Theatrical Tradition* (London: Routledge, 1997), 11–12.

49. There is a convenient introduction to Seneca's trimeters, with references to fuller treatments, in Atze J. Keulen, ed., *L. Annaeus Seneca:* Troades*: Introduction, Text, and Commentary* (Leiden: Brill, 2001), 23–24.

50. Compare and contrast Euripides, *Heracles* 1159–1228, where, upon the arrival of Theseus, the hero covers his head and falls silent, expressing himself only in the sign language of his bloodied hands. See Silvia Montiglio, *Silence in the Land*, 18–19, 225–26, on Heracles' (temporary) silences in this play and in Sophocles' *Trachinian Women*. As she perceptively observes, "Silence and cries are thus homologous, rather than opposite, phenomena, because both signify the collapse of the logos. This is why Heracles' madness, like his agony, is heralded by silence" (225). Connecting Heracles to a post-Homeric "new ideal of heroism, which favors the endurance of the hoplite," Montiglio concludes that "Heracles, and not Odysseus, represents the mythic avatar of the enduring in tragedy. It is Heracles who is celebrated for his strength to face toils. His exceptional ability to endure brings out the hardship of the trials that he endures on the tragic stage, and which are also toils, *mochthoi*" (286). The vocal deformation of Hercules in Seneca can partly be compared to that of the victim of the titular emotion of his treatise *On Anger* (1.1.3–4; 2.35.3–5; 3.4.1–3), itself a model for his vocal portrait of Claudius in the *Apocolocyntosis*, on the connections between which see Osgood, "*Vox* and *Verba*," 347–49.

51. See Chapter 2, on a text's "voice-function."

52. Suetonius, *Nero* 21.3 (who also mentions, in passing at 39.3, a performance as Nauplius) and Dio Cassius, *Roman History* 63.9.4; 63.10.2; 63.22.5–6 (Loeb: vol. 8, pp. 150, 154, 174). Other sources add other titles. See n. 44.

53. On the ancient sources' construction of Nero's endlessly twinned roles as despot and actor, see the brilliant account of Bartsch, *Actors in the Audience*, 36–62.

54. On ancient literature's recurring interest in madness, see generally Ruth Padel, *In and Out of the Mind: Greek Images of the Tragic Self* (Princeton: Princeton University Press, 1992), who notes, "Madness is central to tragedy" (163); Ruth Padel, *Whom Gods Destroy: Elements of Greek and Tragic Madness* (Princeton: Princeton University Press, 1995); Debra Hershkowitz, *The Madness of Epic: Reading Insanity from Homer to Statius* (Oxford: Oxford University Press, 1998); and, for a useful collection of still broader material, Ainsworth O'Brien-Moore, *Madness in Ancient Literature* (Weimar: R. Wagner Sohn, 1924). Each offers sporadic commentary on mad voices.

55. Nero, *frag.* 5 (ed. Courtney), *sub terris tonuisse putes*, which survives only because some wit turned it into a joke about a fellow user of a public toilet. For a cautious attempt to assess Nero's talent on the basis of such fragments, see Henry Bardon, "Les poésies de Néron," *Revue des études latines* 14.1 (1936): 336–49.

56. Aristotle, *Poetics* 4.1449a. The Aristotelian explanation is defended in the standard accounts of tragedy's origins by Albin Lesky, *Greek Tragic Poetry*, trans. Matthew Dillon (New Haven: Yale University Press, 1983), 1–24; it also underpins Pat Easterling's account in the entry on "tragedy, Greek" in Simon Hornblower and Antony Spawforth, eds., *The Oxford Classical Dictionary*, 4th ed. (Oxford: Oxford University Press, 2012), 1493–94.

57. John J. Winkler and Froma I. Zeitlin, eds., *Nothing to Do with Dionysos? Athenian Drama in Its Social Context* (Princeton: Princeton University Press, 1990), who by the question mark they add to their title are already proposing, they explain, that Athenian drama may in fact have had "everything to do with Dionysus" in a broader, social sense (3). Most of their contributors offer readings in a political key, but a noteworthy exception is Jesper Svenbro, "The 'Interior' Voice: On the Invention of Silent Reading," 366–95, who brilliantly suggests that the Greek theater staged something like silent reading, with "passive reading" by the spectators of the "vocal writing" of the actors.

58. Scott Scullion, "'Nothing to Do with Dionysus': Tragedy Misconceived as Ritual," *Classical Quarterly* 52.1 (2002): 102–37.

59. Aristotle, *Poetics* 11.1452b. Jeffrey Walker, "Pathos and Katharsis in 'Aristotelian' Rhetoric: Some Implications," in *Rereading Aristotle's "Rhetoric,"* ed. Alan G. Gross and Arthur E. Walzer (Carbondale: Southern Illinois University

Press, 2000), 74–92, offers a careful "rereading" of the role of *pathos* in Aristotle's poetic, musical, and rhetorical thought, which he connects both to Aristotle's own tentative embrace, in the *Rhetoric*, of the voice (and, more broadly, of *hupokrisis*, "delivery," on which see the following chapter) and to vocal delivery's rapidly expanding importance in subsequent Aristotelian thought, including, famously, that of his pupil and successor, Theophrastus.

60. Mark Payne, "The Understanding Ear: Synaesthesia, Paraesthesia and Talking Animals," in *Synaesthesia and the Ancient Senses*, ed. Shane Butler and Alex Purves (Durham, UK: Acumen, 2013), 47–49, compares these to the animal sounds of comedy. Both, for Payne, reveal "connectivity with nonhuman lives" (52), but "the tragic hero in pain" contrasts with the comic "occasion for metrical, vocal, and, surely, musical exuberance," offering a "thrill" that he calls "an affective high point and an aesthetic low point" (49). This last characterization depends in part on a narrow meaning of "aesthetic" (i.e., one that includes "musical exuberance" but not an irrational "thrill"), but the general contrast relies, to my mind, on confusion of the "alienation" experienced by auditors *in* a tragedy (such as those who are repulsed by the cries of Philoctetes) with the more complex reactions (and, indeed, pleasures) of a theater audience.

61. Kenneth Clark, *The Nude: A Study in Ideal Form* (Princeton: Princeton University Press, 1956), 180, on Myron's *Diskobolos*.

62. Dio Cassius, *Roman History* 63.20.5, 8:168–70 (Loeb).

63. Champlin, *Nero* 59, noting Nero's criticism of his friend Mithridates for a similar stunt.

64. Pseudo-Lucian, *Nero* 3.

65. Katherine E. Welch, *The Roman Amphitheatre: From Its Origins to the Colosseum* (Cambridge: Cambridge University Press, 2007), 151–58.

66. Ovid, *Fasti* 6.797–812. On the iconographic tradition of Heracles *kitharōidos*, see Power, *Culture of Kitharōidia*, 285–93; also n. 77.

67. As G. Karl Galinsky, *The Herakles Theme: The Adaptations of the Hero in Literature from Homer to the Twentieth Century* (Oxford: Blackwell, 1972), 174, aptly puts it, "By a deliberate coincidence of form and content, *Hercules Oetaeus* is every bit as colossal as Herakles himself."

68. Pseudo-Seneca, *Hercules on Oeta* 1128–30. Greek tragedy offers a number of parallels to this Choral report of a noisy approach, such as that of the title character of Sophocles, *Philoctetes* 202–209.

69. The same contrast is in the episode's clear model: Seneca, *Hercules* 569ff., where the Chorus sings of Orpheus (ending by suggesting that if his song had conquered the Underworld, so too can Herculean strength), and Hercules begins by invoking the Sun.

70. As Aristotle recognizes at *Poetics* 4.1449a.

71. See, e.g., Diodorus Siculus 3.67.2. For more on this episode and, more generally, Heracles *mousikos*, see *Lexicon Iconographicum Mythologiae Classicae* 4 "Herakles" II G (1438–82) and III C (1666–73).

72. Edith Hall, "Actor's Song in Tragedy," in *Performance Culture and Athenian Democracy*, ed. Simon Goldhill and Robin Osborne (Cambridge: Cambridge University Press, 1999), 112: "Aeschylean and Euripidean singers are generally the 'others' of the free Greek man in his prime," that is, pre-adult, elderly, female, servile, or foreign.

73. Ibid. Sophocles himself, it seems worth mentioning, reportedly gave up acting because of a weak voice (*Life of Sophocles* 4).

74. Sarah Nooter, *When Heroes Sing: Sophocles and the Shifting Soundscape of Tragedy* (Cambridge: Cambridge University Press, 2012), 73.

75. Cicero, *Tusculan Disputations* 2.20.

76. M.S. Silk, "Heracles and Greek Tragedy," *Greece & Rome* 32.1 (1985): 3–4.

77. Ibid., 4, 19. Hercules' presence throughout ancient performance is also demonstrated by the visual record, on which see Rainer Vollkommer, *Herakles in the Art of Classical Greece* (Oxford: Oxford University Committee for Archaeology, 1988), 61–78.

78. Silk, "Herakles and Greek Tragedy," 9, 16; Euripides, *Heracles* 1412.

79. Silk, "Herakles and Greek Tragedy," 12.

80. Cf. Brooke Holmes, "Euripides' Heracles in the Flesh," *Classical Antiquity* 27.2 (2008), 233 ("Heracles infamous body—the body of brute strength; the body enslaved to impinging forces and uncontrollable appetites; the body that suffers while acting, and is feminized by suffering"), 252 ("the only two tragedies known to have featured Heracles as a protagonist . . . construct his conflicted, tragic identity by joining his boundless strength to the figure of disease, thereby placing his body, with its enormous capacity for inflicting and suffering pain, center stage").

81. Ibid., 235.

82. Ibid., 263 ("If the *tuchē* of Hera strikes a single blow [1393], that blow turns

out to reveal the body as innately vulnerable to misfortunes that disrupt it *from within*"), cf. 273 ("multiple perspectives on the eruption of disorder from within"); 233 ("erupts into visibility"), cf. 270 ("the moment at which the Heracleian body exploded into visibility").

83. As Holmes herself notes, among Heracles' symptoms, "the voice disappears and is reborn as a bellow" (262). She returns, more briefly, to the Euripidean symptomatology of Heracles' madness in Brooke Holmes, *The Symptom and the Subject: The Emergence of the Physical Body in Ancient Greece* (Princeton: Princeton University Press, 2010), 234–45, 242–46, offering the interesting suggestion of a connection to Euripides' popularization of the late fifth century's so-called "new music," which was "particularly well suited to expressing pathos" (235). Again, however, we must keep in mind that Euripides' Heracles did not himself sing.

84. Euripides, *Heracles* 869–70. Nooter, *When Heroes Sing*, 74–75 n. 53, offers suggestive commentary on Heracles' onstage and offstage personas.

85. Elizabeth W. Scharffenberger, "*Deinon Eribremetas*: The Sound and Sense of Aeschylus in Aristophanes' *Frogs*," *Classical World* 100.3 (2007): 232–38.

86. Carl Ruck, "Duality and the Madness of Herakles," *Arethusa* 9.1 (1976): 57.

87. Christina S. Kraus, "Dangerous Supplements: Etymology and Genealogy in Euripides' *Heracles*," *Proceedings of the Cambridge Philological Society* 44 (1999): 137–57.

88. For suggestive commentary on the complex music (and, more broadly, *mousikē*) and musical symbolism of the Chorus and it accompanying *aulos* in Euripides' *Heracles*, see Peter Wilson, "Euripides' Tragic Muse," *Illinois Classical Studies* 24/25 (1999/2000): 433–39.

89. Pseudo-Seneca, *Hercules on Oeta* 1279.

90. Sophocles, *The Trachinian Women* 1275–78.

91. Compare, however, Emma Dillon's category of the "supermusical," by which she designates the propulsive force by which medieval song, exemplified by the motet, reached both beyond words and deeper into language's fiber. In a chapter on "Madness and the Eloquence of Nonsense" in *The Sense of Sound: Musical Meaning in France, 1260–1330* (New York: Oxford University Press, 2012), 129–73, Dillon focuses on rhyme: "At the conjunction of poetry, music, and theory, we arrive at the possibility for hearing rhyme not only as a building block for poetry, but as a sound signifying the etymology of language itself.... If the supermusical seems to disconnect sound from the sense of language, heard another way, it could

be said to make the very essence of language accessible to the ears, and nowhere more so than in its play on rhyme" (170).

92. For a brief exploration of the broader role of embodied acting in Greek drama, see Kostas Valakas, "The Use of the Body by Actors in Tragedy and Satyr-Play," in *Greek and Roman Actors: Aspects of an Ancient Profession*, ed. Pat Easterling and Edith Hall (Cambridge: Cambridge University Press, 2002), 69–92. Commenting on, *inter alia*, *The Trachinian Women* (82-83), Valakas concludes, "The purpose of *nosos* scenes seems to be to represent mythical figures discovering their bodily identity and inner state; this can explain why these scenes are designed to be as strongly based on physical acting and contact as on language and on abstract elements of theatricality" (84). On the full purview of *mousikē*, "that union of song, dance, and word to which the Muses gave their name," see Penelope Murray and Peter Wilson, ed., *Music and the Muses: The Culture of Mousikē in the Classical Athenian City* (Oxford: Oxford University Press, 2004), 1-8.

93. Jimi Hendrix, "Are You Experienced?" The Jimi Hendrix Experience, *Are You Experienced?* (Sony Music Entertainment 88697 6162 2).

94. Charles White, *The Life and Times of Little Richard: The Authorised Biography*, 3rd ed. (London: Omnibus Press, 2003), 228, who dates the comment to 1966.

95. Aristotle, *Problems* 19.43.922a; Pindar, *Pythian Odes* 12.11–21. Plato famously bans the *aulos* from his ideal *polis* in *Republic* 399d. For an extended introduction to the *aulos*, from its mythical invention (and rejection) by Athena to its physical form to its broader cultural resonances, see Peter Wilson, "The *Aulos* in Athens," in *Performance Culture and Athenian Democracy*, ed. Simon Goldhill and Robin Osborne (Cambridge: Cambridge University Press, 1999), 58–95, and, more directly focused on the *aulos* player (*aulētēs*) himself, "The Musicians among the Actors," in *Greek and Roman Actors: Aspects of an Ancient Profession*, ed. Pat Easterling and Edith Hall (Cambridge: Cambridge University Press, 2002), 39–68. Wilson, however, should be read alongside the criticisms and clarifications of Richard P. Martin, "The Pipes Are Brawling: Conceptualizing Musical Performance in Athens," in *The Cultures Within Ancient Greek Culture: Contact, Conflict, Collaboration*, ed. Dougherty and Leslie Kurke (Cambridge: Cambridge University Press, 2003), 153–80.

96. Aristotle, *Politics* 8.1341a24–25.

97. Nicole Loraux, *The Mourning Voice: An Essay on Greek Tragedy* (Ithaca, NY: Cornell University Press, 2002), 48, 72.

98. Loraux, *Mourning Voice*, 82.

99. Plato, *Republic* 3.10.398c–399e; criticized by Aristotle, *Poetics* 8.1342a–b. Loraux, *Mourning Voice*, 60–61.

100. Nietzsche, *Birth of Tragedy*, 21 (Chapter 2).

101. Loraux, *Mourning Voice*, 59.

102. Ibid., 67.

103. Ibid., 54–55.

104. Nietzsche, *Birth of Tragedy*, 81 (Chapter 17).

105. Loraux, *Mourning Voice*, 38–39.

106. On which see the essential study of James I. Porter, *Nietzsche and the Philology of the Future* (Stanford: Stanford University Press, 2000), 127–66.

107. That masks were thought to serve this function is reflected in ancient etymologists' (false) derivation, relayed by Aulus Gellius, *Attic Nights* 5.7, of the Latin *persona* ("mask") from *personare* ("to sound through"). For doubts, see, with earlier bibliography, B. Hunningher, *Acoustics and Acting in the Theatre of Dionysus Eleuthereus* (Amsterdam: Noord-Hollandsche Uitgevers Maatschappij, 1956), 18–19 (320–21).

TRACK FIVE: AMAZING GRACE

1. The manuscript, subsequently lost, also contained the *Letters to Quintus* and some of the *Letters to Brutus*.

2. Petrarch, *Familiares* 24.3.1.

3. Compare the *flebiles voces*, "weepy words," that Seneca the Younger found in another letter to Atticus (lost to us, or perhaps misremembered or invented by Seneca), discussed in *De brevitate vitae* 5. The phrase, though, is almost certainly borrowed from a (doubly) nonepistolary context, for it appears in Cicero's quotation of a noisy lament (probably by Philoctetes) from a lost tragedy, in his dialogue *Tusculan Disputations* 2.33 (cf. 2.55), where he urges against undignified displays of grief. Cleverly redeploying the phrase, Seneca first seems to be pinpointing this very vice in the letter's maudlin tone, but his real target is a single word, *semiliber*, "half-free," a philosophical oxymoron egregious enough, he suggests, to make you cry. (Thanks to Ornella Rossi for pointing out this comparandum to me.)

4. Cicero, *Letters to Atticus* 3.5.

5. The drama is heightened by the tension, in any effort to pronounce the sentence in a sensible way, between *ego*-squashing elision and howling (but

ego-preserving) hiatus. For similar alliteration, see *Letters to Atticus* 3.8.4: *me et meorum malorum maeror et metus de fratre in scribendo impedit.*

6. Petrarch, *Seniles* 16.1.1.

7. On "sweetness" as an aesthetic term in the Middle Ages, see Mary Carruthers, "Sweetness," *Speculum* 81.4 (2006): 999–1013. On Cicero's own terms for such, see A.H. Mamoojee, "'Suavis' and 'Dulcis': A Study of Ciceronian Usage," *Phoenix* 35.3 (1981): 220–36.

8. Text and punctuation from Francesco Petrarca, *Rime, Trionfi, e Poesie latine*, ed. F. Neri et al. (Milan: Riccardo Ricciardi, 1951), 3, including the reading "non che"; translation mine.

9. Petrarch, *Africa* 3.378–81.

10. Petrarch, *Seniles* 16.1.2. On the importance of the discovery to Petrarch's development as a "humanist," see Michael D. Reeve, "Classical Scholarship," in *The Cambridge Companion to Renaissance Humanism*, ed. Jill Kraye (Cambridge: Cambridge University Press, 1996), 20–46.

11. Cicero, *Pro Archia* 19.

12. Ibid. 1.

13. Ibid. 19, borrowed here from L. Laurand, *Études sur le style des discours de Cicéron*, 4th ed. (Amsterdam: Hakkert, 1965), 214, with his indication of *clausulae*, though I would mark these slightly differently. For the *clausulae* of a longer passage from the same speech, see Eduard Fraenkel, *Leseproben aus Reden Ciceros und Catos* (Rome: Edizioni di Storia e Letteratura, 1968), 72–3.

14. John Dugan, "Cicero's Rhetorical Theory," in *The Cambridge Companion to Cicero*, ed. Catherine E.W Steel (Cambridge: Cambridge University Press, 2013), 39.

15. Ronald G. Witt, *In the Footsteps of the Ancients: The Origins of Humanism from Lovato to Bruni* (Leiden: Brill, 2000), 273–74, 509–14: "The frequency of *cursus* in Petrarch's writings appears to vary with the degree of freedom he felt he had to innovate" (273); thus, by Witt's measure, it is "relatively high" in the letters (274), but much lower in, e.g., the *De viris illustribus* (where it characterizes only half of period endings, by his count). As Witt notes, such measures do depend on what you count as *cursus*, and he generously reports the different numbers and conclusions reached by others. However, Giovanni Orlandi, "Metrical and Rhythmical Clausulae in Medieval Latin Prose: Some Aspects and Problems," in *Aspects of the Language of Latin Prose*, ed. Tobias Reinhardt, Michael Lapidge, and J.N. Adams

(Oxford, 2005), 402–403, may partly have solved the riddle of the conflicting data by observing that Petrarch's adherence either to classical *clausulae* or to medieval *cursus* depended on the particular pattern being deployed. For an attempt to decipher Petrarch's own cryptic reference, at *Familiares* 1.6.6, to his "rare" use of "Isocratic reins," see Silvia Rizzo, *Ricerche sul latino umanistico* (Rome: Edizioni di Storia e Letteratura, 2002), 43–53. For the detection of attempted imitation of ancient rhythm in humanist writing a century later, see Terence O. Tunberg, "A Study of *Clausulae* in Selected Works by Lorenzo Valla," *Humanistica Lovaniensia* 41 (1992): 104–33, who helpfully observes, "For this sort of investigation, the question as to whether the humanists correctly understood the rhythmical practice of the ancient authors is really of secondary importance" (109).

16. Jacopo da Bologna, however, proponent of the *suave dolce melodia*, did set one of Petrarch's sonnets to music while he was still alive, and the sonnets were favorites for musical setting in later periods. Petrarch and the great *Ars nova* composer Philippe de Vitry were close friends and mutual admirers.

17. Pietro Bembo, *Prose della volgar lingua* (Venice: Tacuino, 1525).

18. Poliziano, *Letters* 8.16.2; Erasmus, *Ciceronianus* 1160; Quintilian, *The Orator's Education* 10.2.18.

19. Cicero, *Brutus* 65.

20. Cicero, *On the Orator* 3.16.

21. A.E. Douglas, Review of Nello Martinelli, *La rappresentazione dello stile di Crasso e di Antonio nel De oratore*, *Classical Review* 15.1 (1965), 122. Cicero has Atticus question their perfection at *Brutus* 296.

22. Before suggesting that Cicero succeeds in producing a "recognizable echo" of the two men, Nello Martinelli, in *La rappresentazione dello stile di Crasso e di Antonio nel De oratore* (Rome: Centro di Studi Ciceroniani, 1963), 83–84, himself notes that Cicero is simultaneously "painting a portrait of himself" and compares the distance between Crassus and Antonius to that between Cicero's own *Pro Milone* and *In Pisonem*.

23. Cicero, *On the Orator* 3.32.

24. Cicero, *Brutus* 140–42, reading *convertebatur* for *convertebat*, which seems not to have bothered editors, though its emendation is the only good way to account for the dative, which can hardly refer to what follows. Cicero relates the same Demosthenic *dictum* at *De oratore* 3.213 and *Orator* 56.

25. Cicero, *Brutus* 91.

26. Ibid. 92–93.

27. Ibid. 93–94. Cf. 82.

28. Shane Butler, *The Hand of Cicero* (London: Routledge, 2002), 71–84 and generally.

29. Cicero, *Brutus* 19.

30. Cicero, *On Divination* 2.7.

31. Cicero, *On Duty* 2.3.

32. As Valerius Maximus, *Memorable Doings and Sayings* 8.10(ext).1, who also tells the story, confirms.

33. Cicero, *Brutus* 142.

34. Cicero generally avoids even doubly spondaic endings, as L. P. Wilkinson, *Golden Latin Artistry* (Cambridge: Cambridge University Press, 1963), 157, notes, making this triple spondee even more conspicuous.

35. Based on the list provided ibid., 156.

36. Cicero, *Brutus* 313–14. The passage is clearly the source of Plutarch, *Life of Cicero* 3.

37. Cicero, *Brutus* 316.

38. J. C. Davies, "Molon's Influence on Cicero," *Classical Quarterly* 18.2 (1968): 312, arguing that Molon of Rhodes trained Cicero "to minimize the vocal effort of the orator by the increased use of other accompanying devices" (306) and, specifically, to recognize "that an extra 'nuance' of meaning is required in successive phrases in *abundatia* both to maintain the reader's interest and to minimize the need for it to be achieved by a sustained vocal effort on the part of the orator" (307). On Cicero's training, including possible Peripatetic influence on his Rhodian teachers, and its reflection in his later treatises, see Armin Krumbacher, *Die Stimmbildung der Redner im Altertum bis auf die Zeit Quintilians* (Paderborn: Ferdinand Schöningh, 1920), 44–47.

39. Cicero, *Brutus* 303. On *suavis* applied to the voice, see Mamoojee, "'Suavis' and 'Dulcis,'" 232.

40. Cicero, *De oratore* 1.114.

41. Gellius, *Attic Nights* 1.5.2–3. For *mollitudo vocis* as a virtue (i.e., as "vocal flexibility") that is "absolutely necessary for an orator," see *Rhetorica ad Herennium* 3.20.

42. Quintilian, *The Orator's Education* 11.3.8.

43. Gualtiero Calboli, "Oratore senza microfono," in *Ars rhetorica: Antica e*

nuova, ed. Aldo Ceresa-Gastaldo (Genoa: Istituto di Filologia Classica e Medievale, 1983), 23–56.

44. Quintilian, *The Orator's Education* 11.3. Krumbacher, *Die Stimmbildung der Redner*, 51, provides a very helpful table of Ciceronian influences on Quintilian's account. Though based principally on the *Rhetorica ad Herennium*, Jon Hall, "Oratorical Delivery and the Emotions: Theory and Practice," in *A Companion to Roman Rhetoric*, ed. William J. Dominik and Jon Hall (Chichester, UK: Wiley-Blackwell, 2007), 220–24, conveniently outlines the "taxonomy" of vocal delivery in Roman rhetorical theory.

45. Quintilian, *The Orator's Education* 12.10.12.

46. Ibid. 12.10.51.

47. Ibid. 12.10.54–55.

48. On Cicero's use of Demosthenes ("the only choice left" for Cicero, though Isocrates "had probably influenced him more than any Greek orator") against the Neo-Atticists, see Cecil W. Wooten, *Cicero's "Philippics" and Their Demosthenic Model: The Rhetoric of Crisis* (Chapel Hill: University of North Carolina Press, 1983), 48-9. On "Philippics," see Cicero, *Letters to Brutus* 3.4.

49. Pliny the Younger, *Letters* 2.310, who quotes the line in Greek. Ancient sources preserve a remarkable array of details about Demosthenes' voice, especially regarding his vocal training (speaking with pebbles in his mouth, reciting poetry while running, declaiming against the sea), some of which sought to overcome specific defects (like difficulty pronouncing the Greek letter *rhō*). See, e.g., Plutarch's *Life of Demosthenes* 11.1–2, and, for a complete review of the sources, Krumbacher, *Die Stimmbildung der Redner*, 21–29.

50. Valerius Maximus, *Memorable Doings and Sayings* 8.10(ext.).1.

51. Quintilian, *The Orator's Education* 10.1.106.

52. Ibid. 10.1.108.

53. Cicero, *Orator* 104.

54. Longinus, *On the Sublime* 12.3–5 (where Cicero is compared to a spreading fire); Cicero, *Orator* 97; James I. Porter, "Des sons qu'on ne peut entendre: Cicéron, les 'kritikoi' et la tradition du sublime dans la critique littéraire," in *Cicéron et Philodème: La polémique en philosophie*, ed. Clara Auvray-Assayas and Daniel Delattre (Paris: Éditions Rue d'Ulm, 2001), 322, 339. Porter's sublimation of Cicero's scattered pronouncements on voice is dazzling, but he simultaneously reveals somewhat banally phonographic roots of the (vocal) sublime generally,

noting "l'éternel problème de la culture littéraire grecque: comment insuffler de la vie dans la matière sans vie du texte" (341) as well as Cicero's own comment, at *Orator* 130, on the *spiritus* we miss in books (ibid.). In other words, whatever it may become as an object of aesthetic contemplation, the ancient sublime, especially once we include within it the Ciceronian sublime, can also be an effect of the kind of mediatic pragmatism I describe below. To be clear, Porter's "material sublime," here and elsewhere, never neglects mediatic questions, and he too recognizes these as, at least sometimes, primary (indeed, here they are no less than "l'éternal problème"). It is just that I think that the Ciceronian material invites an even stronger shift in emphasis. On the sublimity of rhetorical periods composed like living bodies (the sort for which Cicero was famous), the elements of which, "by the very fact of coming full circle, acquire voice," see Longinus, *On the Sublime* 40.1.

55. Aristotle, *Rhetoric* 3.2.1.1404b; 3.1.5.1404a. Excellent discussion in James I. Porter, "Rhetoric, Aesthetics, and the Voice," in *The Cambridge Companion to Ancient Rhetoric* (Cambridge: Cambridge University Press, 2009), 96–101.

56. For a cautionary introduction to Cicero's own relationship to these *genera dicendi*, see J. G. F Powell, "Cicero's Style," in *The Cambridge Companion to Cicero*, ed. Catherine E. W Steel (Cambridge: Cambridge University Press, 2013), 52-3.

57. Quintilian, *The Orator's Education* 11.3.45; 11.3.61–65; 11.3.151–53.

58. Dionysius of Halicarnassus, *Demosthenes* 53–54.

59. Dionysius of Halicarnassus, *Isocrates* 13.

60. Isocrates, *To Philip* 26; *Panathenaicus* 17.

61. Aristotle, *Rhetoric* 3.12.1413b–1414a. The passage has attracted an extensive scholarly attention, but see especially Neil O'Sullivan, *Alcidamas, Aristophanes and the Beginnings of Greek Stylistic Theory* (Stuttgart: Franz Steiner, 1992), 42–62, and Richard Graff, "Reading and the 'Written Style' in Aristotle's *Rhetoric*," *Rhetoric Society Quarterly* 31.4 (2001): 19–44.

62. Ibid. 3.12.1413b (trans. Roberts).

63. Seneca the Elder seems alone among ancient writers in expressing regret at not hearing the *viva vox* of Cicero (he just barely could have done so, he says, if civil war had not delayed his boyhood arrival in Rome), but the very phrase "living voice" was shorthand for a platitude (*ut vulgo dicitur*, as Seneca marks it) usually invoked in the breach: see A. Otto, *Die Sprichwörter und sprichwörtlichen Redensarten der Römer: Gesammelt und Erklärt* (Leipzig: B. G. Teubner, 1890), 378. To be sure, Seneca' s contemporaries still mourned Cicero's death as a collective calamity

for Latin eloquence; his *Suasoriae* 6 and 7 collect some of their elegant (and even poetic) groans of grief. But none of them really thought Cicero's voice had died with him, as is clear from the second of these (which closes the collection), in which the gathered guests were invited to declaim on an imagined offer by Antony to spare Cicero's life—if he would but burn his books of speeches.

64. Quintilian, *The Orator's Education* 10.1.112. Petrarch, *Seniles* 16.1.2.

65. On Cicero's constructed persona in the *Brutus*, including vis-à-vis his title character, see John Dugan, *Making a New Man: Ciceronian Self-Fashioning in the Rhetorical Works* (Oxford: Oxford University Press, 2005), 172–250.

66. Seneca the Elder, *Suasoriae* 6.26.

67. That is, *Secretum (meum)*, in Petrarch's own abbreviation (26, ed. Carrara) of its full title, *De secreto conflictu curarum mearum*. The dialogue begins with Petrarch in conversation with Truth herself, who instructs Augustine to take over: "Let a human voice (*humana vox*) strike the ear of a mortal man; he will submit to this more patiently. I shall remain here, however, so that whatever he hears from you he can regard as coming from me" (ibid.).

68. Francesco Di Capua, "Il ritmo prosaico in S. Agostino," in *Miscellanea agostiniana* (Rome: Typografia Poliglotta Vaticana, 1931), 2:678, remarking, "Qui l'autore non declama, ma piange." Note that this avoidance is at odds even with Augustine's own ordinary compositional practice, which he himself describes (*De doctrina Christiana* 4.41) as not neglectful of prose rhythm (*non praetermitto istos numeros clausularum*). Indeed, as Adolf Primmer, "The Function of the *Genera Dicendi* in *De Doctrina Christiana* 4," in "*De Doctrina Christiana*": *A Classic of Western Culture*, ed. Duane W. H. Arnold and Pamela Bright (Notre Dame, IN: University of Notre Dame Press, 1995), 82, notes, Augustine's discussion in that treatise of Ciceronian style is itself structured rhythmically.

69. Augustine, *Confessions* 10.33, on the complexities and influence of which, see Emma Dillon, *The Sense of Sound: Musical Meaning in France, 1260–1330* (New York: Oxford University Press, 2012), 38–43. For a broader selection of Augustine's conflicted pronouncements on both musical and rhetorical sound, see James Anderson Winn, *Unsuspected Eloquence: A History of the Relations Between Poetry and Music* (New Haven: Yale University Press, 1981), 43–55.

70. Catherine Conybeare, "Beyond Word and Image: Aural Patterning in Augustine's *Confessions*," in *Envisioning Experience in Late Antiquity and the Middle Ages: Dynamic Patterns in Texts and Images*, ed. Giselle de Nie and Thomas F. X.

Noble (Farnham, UK: Ashgate, 2012), 157, 160, 164.

71. Augustine, *Confessions* 8.12.

72. Ibid. 3.4.

73. My translation follows the text as rightly emended by O'Donnell, after Löfstedt.

74. Augustine, *De doctrina Christiana* 4.41. Jerome, *Letters* 22.30, to Eustochium.

75. For an illuminating review of the *Confessions* as "an aural history of conversion" (11), describing the "voices of many kinds" (*Confessions* 13.1) that incrementally had led its author to the very condition of being able to hear the voice of God himself (not just in the garden's disembodied voice, but also in the text he picks up and reads), see William North, "Hearing Voices in Late Antiquity: An Aural Approach to Augustine's *Confessions*," in Jason Glen, ed., *The Middle Ages in Texts and Texture: Reflections on Medieval Sources* (Toronto: University of Toronto Press, 2011), 7–19.

76. Augustine *Confessions* 10.31. For the translation, with a beautiful exploration of the theme of the pulse in the *Confessions* and its relation to Augustine's interests in both meter and medicine, see Marjorie O'Rourke Boyle, "Augustine's Heartbeat: From Time to Eternity," *Viator* 38.1 (2007): 19–43. As she notes regarding the major Greek theorist of the pulse, Herophilus, whose theories Augustine knew indirectly, "Augustine transfers Herophilus's pulsation from classical Greek quantitative meter to late Roman accentual meter, or word-stress. The trochee [–⏑] is exemplary, the very beat that Augustine and Monica tap at the climax of their meditation" (32) at Ostia, when he and his mother together reach for God *toto ictu cordis*, "with a whole heartbeat" (*Confessions* 9.24).

77. Marshall McLuhan, *Understanding Media: The Extensions of Man* (New York: McGraw-Hill, 1964).

78. Angelo Poliziano, *Letters* 1.1.2.

79. Michel de Certeau, *The Practice of Everyday Life*, trans. Steven Rendall (Berkeley: University of California Press, 1984), 161, where the sentence is italicized.

Discography

Hendrix, Jimi. "Are You Experienced?" The Jimi Hendrix Experience, *Are You Experienced?* Sony Music Entertainment 88697 6162 2. ℗© 2010 Experience Hendrix, LLC. Lyrics © 1967, 2010 Experience Hendrix, LLC.

Mozart, Wolfgang Amadeus. *Don Giovanni.* The New Philharmonia Orchestra and Chorus. Conducted by Otto Klemperer. With Nicolai Ghiaurov, Franz Crass, Claire Watson, Nicolaï Gedda, Christa Ludwig, Walter Berry, Mirella Freni, and Paulo Montarsolo. Angel Records SDL 3700 (1–4). ℗ 1966 Angel Records.

Page, Jimmy, and Robert Plant. "Stairway to Heaven." *Led Zeppelin IV.* Atlantic Recording Corporation 82638-2. © 1971 Atlantic Recording Corporation. Lyrics © 1971 Superhype Publishing, Inc. Originally released as Atlantic 7208 on November 8, 1971.

Parton, Dolly. *Halos & Horns.* Sugar Hill Records 3946. ℗© 2002 Dolly Parton.

Puccini, Giacomo. *Tosca.* Orchestra and Chorus of La Scala, Milan. Conducted by Victor de Sabata. With Maria Callas, Giuseppe di Stefano, Tito Gobbi. EMI Records SLS 825. ℗ 1953 EMI Records.

———. *Tosca.* Orchestre de la Société des Concerts du Conservatoire. Conducted by Georges Prêtre. With Maria Callas, Carlo Bergonzi, and Tito Gobbi. His Master's Voice SAN 149–50. ℗ 1965 The Gramophone Company.

Trask, Stephen (music and lyrics) and John Cameron Mitchell (vocals). "The Origin of Love." *Hedwig and the Angry Inch.* Original Cast Recording. Music and lyrics by Stephen Trask. Text by John Cameron Mitchell. Atlantic Recording Corporation 83160-2. ℗© 1999 Atlantic Recording Corporation and WEA International, Inc.

Bibliography

Most translations from Greek and Latin in this book are my own. One major exception is the Aristotelian corpus, generally quoted from *The Complete Works of Aristotle: The Revised Oxford Translation*, edited by Jonathan Barnes (Princeton: Princeton University Press, 1984), with the name of the specific translator in parentheses following the citation. Other exceptions are indicated in the notes.

Abbate, Carolyn. *In Search of Opera*. Princeton: Princeton University Press, 2001.

———. *Unsung Voices: Opera and Musical Narrative in the Nineteenth Century*. Princeton: Princeton University Press, 1991.

Adorno, Theodor W. *Essays on Music*. Edited by Richard D. Leppert, translated by Susan H. Gillespie. Berkeley: University of California Press, 2002.

Ahl, Frederick. *Metaformations: Soundplay and Wordplay in Ovid and Other Classical Poets*. Ithaca, NY: Cornell University Press, 1985.

Allen, W. Sidney. *Accent and Rhythm. Prosodic Features of Latin and Greek: A Study in Theory and Reconstruction*. Cambridge: Cambridge University Press, 1973.

———. *Vox Graeca: A Guide to the Pronunciation of Classical Greek*. 3rd ed. Cambridge: Cambridge University Press, 1987.

Anderson, Graham. *Lucian: Theme and Variation in the Second Sophistic*. Leiden: Brill Archive, 1976.

———. *Philostratus: Biography and Belles Lettres in the Second Century A.D.* London: Croom Helm, 1986.

Anderson, Warren D. *Music and Musicians in Ancient Greece*. Ithaca, NY: Cornell University Press, 1994.

Anderson, William S., ed. *Ovid's "Metamorphoses": Books 1–5*. Norman: University of Oklahoma Press, 1997.

Aristotle. *The Complete Works of Aristotle: The Revised Oxford Translation*. Edited by Jonathan Barnes. Princeton: Princeton University Press, 1984.

——. *Organon. I. Catégories. II. De L'interprétation*. Translated by J. Tricot. Paris: Vrin, 1946.

——. *The Works of Aristotle*. Edited by W.D. Ross. Vol. 1. Oxford: Oxford University Press, 1928.

Austin, Roland G. "Virgil and the Sibyl." *Classical Quarterly* 21.2 (1927): 100–105.

Ax, W. "Zum de voce-Kapitel der römischen Grammatik: Eine Antwort auf Dirk M. Schenkeveld und Wilfried Stroh." In *Grammatical Theory and Philosophy of Language in Antiquity*, edited by Pierre Swiggers and Alfons Wouters, 19–121. Leuven: Peeters, 2002.

Barchiesi, Alessandro. "Music for Monsters: Ovid's Metamorphoses, Bucolic Evolution, and Bucolic Criticism." In *Brill's Companion to Greek and Latin Pastoral*, edited by Marco Fantuzzi and Theodore Papanghelis, 403–26. Leiden: Brill, 2006.

——. "Narrative Technique and Narratology in the Metamorphoses." In *The Cambridge Companion to Ovid*, edited by Philip R. Hardie, 180–99. Cambridge: Cambridge University Press, 2002.

——. *Speaking Volumes: Narrative and Intertext in Ovid and Other Latin Poets*. Translated by Matt Fox and Simone Marchesi. London: Duckworth, 2001.

Bardon, Henry. *Les empereurs et le lettres latines d'Auguste à Hadrien*. 2nd ed., rev. Paris: Les Belles Lettres, 1968.

——. "Les poésies de Néron." *Revue des Études Latines* 14.1 (1936): 336–49.

Barker, Andrew, ed. *Greek Musical Writings*. Vol. 2. Cambridge: Cambridge University Press, 1989.

——. "Phōnaskia for Singers and Orators: The Care and Training of the Voice in the Roman Empire." In *La musica nell'Impero romano: Testimonianze teoriche e scoperte archeologiche*, edited by Eleonora Rocconi, 11–20. Pavia: Pavia University Press, 2010.

Barthes, Roland. "Écoute." In *L'obvie et l'obtus: Essais critiques III*, 217–30. Paris: Éditions du Seuil, 1982.

——. "The Grain of the Voice." In *Image, Music, Text*, translated by Stephen Heath, 179–89. New York: Hill and Wang, 1977.

——. "Listening." In *The Responsibility of Forms: Critical Essays on Music, Art, and Representation*, translated by Richard Howard, 245–60. New York: Hill and Wang, 1985.

——. *The Rustle of Language*. Berkeley: University of California Press, 1986.

Bartsch, Shadi. *Actors in the Audience: Theatricality and Doublespeak from Nero to Hadrian*. Cambridge, MA: Harvard University Press, 1994.

——. *The Mirror of the Self: Sexuality, Self-Knowledge, and the Gaze in the Early Roman Empire*. Chicago: University of Chicago Press, 2006.

Battezzato, Luigi. "Metre and Music." In *The Cambridge Companion to Greek Lyric*, edited by Felix Budelmann, 130–46. Cambridge: Cambridge University Press, 2009.

Beekes, R. S. P. *Etymological Dictionary of Greek*. Leiden: Brill, 2010.

Bembo, Pietro. *Prose . . . della volgar lingua*. Venice: Tacuino, 1525.

Bergk, Theodor, ed. *Poetae Lyrici Graeci*. 4th ed. Vol. 3. Leipzig: B. G. Teubner, 1882.

Bettini, Maurizio. *Voci: Antropologia sonora del mondo antico*. Turin: G. Einaudi, 2008.

Bicknell, C. D. "Some Vases in the Lewis Collection." *Journal of Hellenic Studies* 41 (1921): 222–31.

Billerbeck, Margarethe. "Hercules Bound: A Note on Suetonius, *Nero* 21.3." *American Journal of Philology* 102.1 (1981): 54–57.

Bologna, Corrado. *Flatus vocis: Metafisica e antropologia della voce*. Bologna: Il Mulino, 1992.

Bottini, Angelo, and Pier Giovanni Guzzo. "Orfeo e le Sirene al Getty Museum." *Ostraka: Rivista di antichità* 2.1 (1993): 43–52.

Bowden, Hugh. *Classical Athens and the Delphic Oracle: Divination and Democracy*. Cambridge: Cambridge University Press, 2005.

Boyle, A. J. *Tragic Seneca: An Essay in the Theatrical Tradition*. London: Routledge, 1997.

Bradley, K. R. *Suetonius' Life of Nero: An Historical Commentary*. Brussels: Latomus, 1978.

Breed, Brian W. *Pastoral Inscriptions: Reading and Writing Virgil's "Eclogues."* London: Duckworth, 2006.

Brenkman, John. "Narcissus in the Text." *Georgia Review* 30 (1976): 293–327.

Budelmann, Felix. "Anacreon and the Anacreontea." In *The Cambridge Companion to Greek Lyric*, edited by Felix Budelmann, 227–39. Cambridge: Cambridge University Press, 2009.

Burkert, Walter. *Homo Necans: The Anthropology of Ancient Greek Sacrificial Ritual and Myth*. Berkeley: University of California Press, 1983.

Burmann, P., ed. *Publii Ovidii Nasonis Opera Omnia*. Vol. 2. Amsterdam: Apud R. et J. Westenios et G. Smith, 1727.

Burnyeat, M. F. "Postscript on Silent Reading." *Classical Quarterly* 47.1 (1997): 74–76.

Butler, Shane. "Beyond Narcissus." In *Synaesthesia and the Ancient Senses*, edited by Shane Butler and Alex Purves, 185–200. Durham, UK: Acumen Publishing, 2013.

——. *The Hand of Cicero*. London: Routledge, 2002.

——. *The Matter of the Page: Essays in Search of Ancient and Medieval Authors*. Madison: University of Wisconsin Press, 2011.

Calboli, Gualtiero. "Oratore senza microfono." In *Ars rhetorica: Antica e nuova*, edited by Aldo Ceresa-Gastaldo, 23–56. Genoa: Istituto di Filologia Classica e Medievale, 1983.

Carruthers, Mary. "Sweetness." *Speculum* 81.4 (2006): 999–1013.

Carson, Anne. "The Gender of Sound." In *Glass, Irony, and God*, 119–37. New York: New Directions, 1995.

Cavarero, Adriana. *For More than One Voice: Toward a Philosophy of Vocal Expression*. Translated by Paul A. Kottman. Palo Alto: Stanford University Press, 2005.

Certeau, Michel de. *The Practice of Everyday Life*. Translated by Steven Rendall. Berkeley: University of California Press, 1984.

——. "Vocal Utopias: Glossolalias." Translated by Daniel Rosenberg. *Representations* 56 (1996): 29–47.

Champlin, Edward. *Nero*. Cambridge, MA: Harvard University Press, 2003.

Clark, Kenneth. *The Nude: A Study in Ideal Form*. Princeton: Princeton University Press, 1956.

Colish, Marcia L. *The Stoic Tradition from Antiquity to the Early Middle Ages*. Leiden: E. J. Brill, 1985.

Connor, Steven. *Beyond Words: Sobs, Hums, Stutters and Other Vocalizations*. London: Reaktion Books, 2014.

——. *Dumbstruck: A Cultural History of Ventriloquism*. Oxford: Oxford University Press, 2000.

Conte, Gian Biagio. *Latin Literature: A History*. Translated by Joseph B. Solodow. Baltimore: Johns Hopkins University Press, 1994.

Conybeare, Catherine. "Beyond Word and Image: Aural Patterning in Augustine's *Confessions*." In *Envisioning Experience in Late Antiquity and the Middle Ages:*

Dynamic Patterns in Texts and Images, edited by Giselle de Nie and Thomas F. X. Noble, 143–64. Farnham, UK: Ashgate, 2012.

Dahlhaus, Carl. *The Idea of Absolute Music*. Chicago: University of Chicago Press, 1989.

Damon, Phillip. *Modes of Analogy in Ancient and Medieval Verse*. Berkeley: University of California Press, 1961.

D'Angour, Armand. *The Greeks and the New: Novelty in Ancient Greek Imagination and Experience*. Cambridge: Cambridge University Press, 2011.

Davies, J. C. "Molon's Influence on Cicero." *Classical Quarterly* 18.2 (1968): 303–14.

De Martino, Francesco. "La voce degli autori." In *Lo spettacolo delle voci*, edited by Francesco De Martino and Alan H. Sommerstein, 17–59. Bari: Levante, 1995.

Derrida, Jacques. "Aphorism Countertime." In *Acts of Literature*, edited by Derek Attridge, 416–33. London: Routledge, 1992.

——. *De la grammatologie*. Paris: Éditions de Minuit, 1967.

——. "Freud and the Scene of Writing." In *Writing and Difference*, translated by Alan Bass, 196–231. Chicago: University of Chicago Press, 1978.

——. *Of Grammatology*. Translated by Gayatri Chakravorty Spivak. Baltimore: Johns Hopkins University Press, 1976.

Di Capua, Francesco. "Il ritmo prosaico in S. Agostino." In *Miscellanea Agostiniana*, 2:607–64. Rome: Typografia Poliglotta Vaticana, 1931.

Dickey, Eleanor. *Ancient Greek Scholarship*. New York: Oxford University Press, 2006.

Dillon, Emma. *The Sense of Sound: Musical Meaning in France, 1260–1330*. New York: Oxford University Press, 2012.

Dolar, Mladen. *A Voice and Nothing More*. Cambridge, MA: MIT Press, 2006.

Douglas, A. E. Review of Nello Martinelli, *La rappresentazione dello stile di Crasso e di Antonio nel De oratore*. *Classical Review* 15.1 (1965): 122.

Drucker, Johanna, and Emily McVarish. *Graphic! Design History*. Upper Saddle River, NJ: Pearson Prentice Hall, 2009.

Dugan, John. "Cicero's Rhetorical Theory." In *The Cambridge Companion to Cicero*, edited by Catherine E. W. Steel, 25–40. Cambridge: Cambridge University Press, 2013.

——. *Making a New Man: Ciceronian Self-Fashioning in the Rhetorical Works*. Oxford: Oxford University Press, 2005.

Dupont, Florence. *The Invention of Literature: From Greek Intoxication to the Latin*

Book. Translated by Janet Lloyd. Baltimore: Johns Hopkins University Press, 1999.

Easterling, Pat. "Actors and Voices: Reading Between the Lines in Aeschines and Demosthenes." In *Performance Culture and Athenian Democracy*, edited by Simon Goldhill and Robin Osborne, 154–66. Cambridge: Cambridge University Press, 1999.

Edison, Thomas A. "Improvement in Phonograph or Speaking Machine," February 19, 1878. http://edison.rutgers.edu/patent1.htm.

Eidsheim, Nina Sun. "Sensing Voice: Materiality and the Lived Body in Singing and Listening." *Senses & Society* 6.2 (2011): 133–55.

Eijk, Philip J. van der. *Medicine and Philosophy in Classical Antiquity: Doctors and Philosophers on Nature, Soul, Health and Disease*. Cambridge: Cambridge University Press, 2005.

Empson, William. *Seven Types of Ambiguity*. 3rd ed. Harmondsworth, UK: Penguin Books, 1961.

Enterline, Lynn. *The Rhetoric of the Body from Ovid to Shakespeare*. Cambridge: Cambridge University Press, 2000.

Erlmann, Veit. *Reason and Resonance: A History of Modern Aurality*. New York: Zone Books, 2010.

Evans, Joseph Claude. *Strategies of Deconstruction: Derrida and the Myth of the Voice*. Minneapolis: University of Minnesota Press, 1991.

Ferruzza, Maria Lucia. *Ancient Terracottas from South Italy and Sicily in the J. Paul Getty Museum*. Malibu, CA: J. Paul Getty Museum, in press.

Fontenrose, Joseph. *The Delphic Oracle: Its Responses and Operations, with a Catalogue of Responses*. Berkeley: University of California Press, 1978.

Ford, Andrew L. "Dionysos' Many Names in Aristophanes' Frogs." In *A Different God? Dionysos and Ancient Polytheism*, edited by Renate Schlesier, 343–55. Berlin: De Gruyter, 2012.

Foucault, Michel. "What Is an Author?" In *Language, Counter-Memory, Practice: Selected Essays and Interviews*, edited by Donald F. Bouchard, translated by Donald F. Bouchard and Sherry Simon, 113–38. Ithaca, NY: Cornell University Press, 1977.

Fraenkel, Eduard. *Leseproben aus Reden Ciceros und Catos*. Rome: Edizioni di Storia e Letteratura, 1968.

Frazer, R. M. "Nero the Singing Animal." *Arethusa* 4.2 (1971): 215–18.

Frenk-Oczlon, Gertraud, and August Fenk. "Musical Pitch in Nonsense Syllables: Correlations with the Vowel System and Evolutionary Perspectives." In *Proceedings of the Seventh Triennial Conference of the European Society for the Cognitive Sciences of Music*, edited by J. Louhivou et al., 110–13. Jyväskylä, Finland: University of Jyväskylä, 2009.

Freud, Sigmund. *New Introductory Lectures on Psycho-analysis*. Translated by James Strachey. London: Vintage, 2001.

Galinsky, G. Karl. *The Herakles Theme: The Adaptations of the Hero in Literature from Homer to the Twentieth Century*. Oxford: Blackwell, 1972.

Garrett, Stewart. *Reading Voices: Literature and the Phonotext*. Berkeley: University of California Press, 1990.

Gavrilov, A. K. "Techniques of Reading in Classical Antiquity." *Classical Quarterly* 47.1 (1997): 56–73.

Genette, Gérard. *Mimologics*. Translated by Thaïs E. Morgan. Lincoln: University of Nebraska Press, 1995.

Gerber, Douglas E. *A Companion to the Greek Lyric Poets*. Leiden: Brill, 1997.

Gildenhard, Ingo, and Andrew Zissos. "Ovid's Narcissus (Met. 3.339–510): Echoes of Oedipus." *American Journal of Philology* 121.1 (2000): 129–47.

Gitelman, Lisa. *Scripts, Grooves, and Writing Machines: Representing Technology in the Edison Era*. Stanford: Stanford University Press, 1999.

Golden, Mark. "Baby Talk and Child Language in Ancient Greece." In *Lo spettacolo delle voci*, edited by Francesco De Martino and Alan H. Sommerstein, 11–33. Bari: Levante, 1995.

Goldhill, Simon. "The Language of Tragedy: Rhetoric and Communication." In *The Cambridge Companion to Greek Tragedy*, edited by P. E. Easterling, 127–50. Cambridge: Cambridge University Press, 1997.

Graff, Richard. "Reading and the 'Written Style' in Aristotle's Rhetoric." *Rhetoric Society Quarterly* 31.4 (2001): 19–44.

Grover-Friedlander, Michal. *Operatic Afterlives*. New York: Zone Books, 2011.

———. *Vocal Apparitions: The Attraction of Cinema to Opera*. Princeton: Princeton University Press, 2005.

Gyles, Mary Francis. "Nero: Qualis Artifex?" *Classical Journal* 57.5 (1962): 193–200.

———. "Nero Fiddled While Rome Burned." *Classical Journal* 42.4 (1947): 211–17.

Habinek, Thomas. "Singing, Speaking, Making, Writing: Classical Alternatives to

Literature and Literary Studies." *Stanford Humanities Review* 6.1 (1998): 65–75.

——. *The World of Roman Song: From Ritualized Speech to Social Order*. Baltimore: Johns Hopkins University Press, 2005.

Hagel, Stefan. *Ancient Greek Music: A New Technical History*. Cambridge: Cambridge University Press, 2010.

Hall, Edith. "Actor's Song in Tragedy." In *Performance Culture and Athenian Democracy*, edited by Simon Goldhill and Robin Osborne, 96–122. Cambridge: Cambridge University Press, 1999.

——. *Inventing the Barbarian: Greek Self-Definition Through Tragedy*. Oxford: Oxford University Press, 1989.

——. "The Singing Actors of Antiquity." In *Greek and Roman Actors: Aspects of an Ancient Profession*, edited by Pat Easterling and Edith Hall, 4–38. Cambridge: Cambridge University Press, 2002.

Hall, Jon. "Oratorical Delivery and the Emotions: Theory and Practice." In *A Companion to Roman Rhetoric*, edited by William J. Dominik and Jon Hall, 218–34. Chichester: Wiley-Blackwell, 2007.

Hanslick, Eduard. *The Beautiful in Music: A Contribution to the Revival of Musical Aesthetics*. London: Novello and Co., 1891.

Hardie, Philip. *Ovid's Poetics of Illusion*. Cambridge: Cambridge University Press, 2002.

——. *Rumour and Renown: Representations of* Fama *in Western Literature*. Cambridge: Cambridge University Press, 2012.

Harley, Trevor A. *The Psychology of Language: From Data to Theory*. 3rd ed. New York: Psychology Press, 2008.

Harmon, A. M., trans. *Lucian, with an English Translation*. Vol. 4. London: Heinemann, 1913.

Harris, Roy. *The Origin of Writing*. London: Duckworth, 1986.

Hegel, Georg Wilhelm Friedrich. *Lectures on the Philosophy of History*. Translated by J. Sibree. London: Henry G. Bohn, 1857.

Heller-Roazen, Daniel. *Echolalias: On the Forgetting of Language*. New York: Zone Books, 2005.

Hemelt, Francis J. "Points of Resemblance in the Verse of Tennyson and Theocritus." *Modern Language Notes* 18.4 (1903): 115–17.

Henderson, W. J. "Performance of Puccini's 'Tosca' at the Metropolitan Opera House." *The New York Times*, February 10, 1901.

Heraeus, W. "Die Sprache der römische Kinderstube." *Archiv für lateinische Lexikographie und Grammatik* 13 (1904): 149–72.

Hershkowitz, Debra. *The Madness of Epic: Reading Insanity from Homer to Statius.* Oxford: Oxford University Press, 1998.

Holford-Strevens, Leofranc. *Aulus Gellius: An Antonine Scholar and His Achievement.* Rev. ed. Oxford: Oxford University Press, 2003.

Hollander, John. "Breaking Into Song: Some Notes on Refrain." In *Lyric Poetry: Beyond New Criticism*, edited by Chaviva Hošek and Patricia A. Parker, 73–89. Ithaca, NY: Cornell University Press, 1985.

——. *The Figure of Echo: A Mode of Allusion in Milton and After.* Berkeley: University of California Press, 1981.

Holmes, Brooke. "Euripides' Heracles in the Flesh." *Classical Antiquity* 27.2 (2008): 231–81.

——. *The Symptom and the Subject: The Emergence of the Physical Body in Ancient Greece.* Princeton: Princeton University Press, 2010.

Hornblower, Simon, and Antony Spawforth, eds. *The Oxford Classical Dictionary.* 4th ed. Oxford: Oxford University Press, 2012.

Huck, William. "Tosca: Four Callas Toscas." *Opera Quarterly* 2.3 (1984): 175–78.

Hunningher, B. *Acoustics and Acting in the Theatre of Dionysus Eleuthereus.* Amsterdam: Noord-Hollandsche Uitgevers Maatschappij, 1956.

Irvine, Martin. *The Making of Textual Culture: "Grammatica" and Literary Theory, 350–1100.* Cambridge: Cambridge University Press, 2006.

Jakobson, Roman. "Why 'Mama' and 'Papa'?" In *Selected Writings*, edited by Stephen Rudy, 538–45. The Hague: Mouton, 1962.

Jakobson, Roman, and Morris Halle. *Fundamentals of Language.* 4th ed. The Hague: Mouton, 1980.

Janan, Micaela Wakil. *Reflections in a Serpent's Eye: Thebes in Ovid's "Metamorphoses."* Oxford: Oxford University Press, 2009.

Johansen, Flemming. *Greek Portraits: Ny Carlsberg Glyptotek.* Copenhagen: Ny Carlsberg Glyptotek, 1992.

Johnson, Edward H. "A Wonderful Invention: Speech Capable of Indefinite Repetition from Automatic Records." *Scientific American* 37.20 (November 17, 1877): 304.

Johnson, William A. *Readers and Reading Culture in the High Roman Empire: A Study of Elite Communities.* New York: Oxford University Press, 2010.

——. "Toward a Sociology of Reading in Classical Antiquity." *American Journal of Philology* 121.4 (2000): 593–627.

Kafka, Franz. "The Silence of the Sirens." In *Homer: A Collection of Critical Essays*, edited by George Steiner and Robert Fagles, 98–99. Englewood Cliffs, NJ: Prentice-Hall, 1962.

Kaimio, Maarit. *Characterization of Sound in Early Greek Literature*. Helsinki: Societas Scientiarum Fennica, 1977.

Keil, Heinrich, ed. *Grammatici Latini*. 7 vols. Leipzig: B. G. Teubner, 1857–80.

Keulen, Atze J., ed. *L. Annaeus Seneca, "Troades": Introduction, Text, and Commentary*. Leiden: Brill, 2001.

Kittler, Friedrich A. *Gramophone, Film, Typewriter*. Translated by Geoffrey Winthrop-Young and Michael Wutz. Stanford: Stanford University Press, 1999.

Knox, Bernard M. W. "Silent Reading in Antiquity." *Greek, Roman, and Byzantine Studies* 9.4 (1968): 421–35.

Knox, Peter E. *Ovid's "Metamorphoses" and the Traditions of Augustan Poetry*. Cambridge: Cambridge Philological Society, 1986.

Kraus, Christina S. "Dangerous Supplements: Etymology and Genealogy in Euripides' *Heracles*." *Proceedings of the Cambridge Philological Society* 44 (1999): 137–57.

Kretzmann, Norman. "Aristotle on Spoken Sound Significant by Convention." In *Ancient Logic and Its Modern Interpretations*, edited by John Corcoran, 3–21. Dordrecht: Reidel, 1974.

Kristeva, Julia. *Revolution in Poetic Language*. Translated by Margaret Waller. New York: Columbia University Press, 1984.

Krumbacher, Armin. *Die Stimmbildung der Redner im Altertum bis auf die Zeit Quintilians*. Paderborn: Ferdinand Schöningh, 1920.

Lacan, Jacques. *The Four Fundamental Concepts of Psychoanalysis*. Translated by Alan Sheridan. New York: Norton, 1981.

——. "The Instance of the Letter in the Unconscious, or Reason Since Freud." In *Écrits*, translated by Bruce Fink, 412–41. New York: Norton, 2006.

——. "The Mirror Stage as Formative of the *I* Function as Revealed in Psychoanalytic Experience." In *Écrits*, translated by Bruce Fink, 75–81. New York: W. W. Norton, 2006.

——. "Seminar 14 (20 April 1966): Summary of Crucial Problems; Jouissance." In *Seminar XIII: The Object of Psychoanalysis*, translated by Cormac Gallagher,

151–62, n.d. http://www.lacaninireland.com/web/published-works/seminars/.

Lachenaud, Guy. *Les routes de la voix: L'antiquité grecque et le mystère de la voix.* Paris: Les Belles Lettres, 2013.

Laurand, L. *Études sur le style des discours de Cicéron.* 4th ed. Amsterdam: Hakkert, 1965.

Leach, Elizabeth Eva. *Sung Birds: Music, Nature, and Poetry in the Later Middle Ages.* Ithaca, NY: Cornell University Press, 2007.

Lesky, Albin. *Greek Tragic Poetry.* Translated by Matthew Dillon. New Haven: Yale University Press, 1983.

LeVen, Pauline A. "The Colors of Sound: *Poikilia* and Its Aesthetic Contexts." *Greek and Roman Musical Studies* 1 (2013): 229–42.

Loewenstein, Joseph. *Responsive Readings: Versions of Echo in Pastoral, Epic, and the Jonsonian Masque.* New Haven: Yale University Press, 1984.

Loraux, Nicole. *The Mourning Voice: An Essay on Greek Tragedy.* Ithaca, NY: Cornell University Press, 2002.

Lowrie, Michèle. *Writing, Performance, and Authority in Augustan Rome.* Oxford: Oxford University Press, 2009.

Macleod, M.D., ed. *Lucian, with an English Translation.* Vol. 8. London: Heinemann, 1967.

Mamoojee, A.H. "'Suavis' and 'Dulcis': A Study of Ciceronian Usage." *Phoenix* 35.3 (1981): 220–36.

Manetti, Giovanni. *Le teorie del segno nell'antichità classica.* Milan: Bompiani, 1987.

———. "Semanticità, articolazione, scrivibilità: Gli spazi di confine tra l'uomo e l'animale nella Grecia antica." In *Ta zōia. L'espai a Grècia II: Els animals i l'espai,* edited by Montserrat Jufresa and Montserrat Reig, 13–26. Tarragona: Institut Català d'Arqueologica Clàssica, 2011.

Martin, Richard P. "The Pipes Are Brawling: Conceptualizing Musical Performance in Athens." In *The Cultures Within Ancient Greek Culture: Contact, Conflict, Collaboration,* edited by Dougherty and Leslie Kurke, 153–80. Cambridge: Cambridge University Press, 2003.

Martinelli, Nello. *La rappresentazione dello stile di Crasso e di Antonio nel De oratore.* Rome: Centro di Studi Ciceroniani, 1963.

Mathiesen, Thomas J. *Apollo's Lyre: Greek Music and Music Theory in Antiquity and the Middle Ages.* Lincoln: University of Nebraska Press, 1999.

McLuhan, Marshall. *The Gutenberg Galaxy: The Making of Typographic Man*. Toronto: University of Toronto Press, 1962.

——. *Understanding Media: The Extensions of Man*. New York: McGraw-Hill, 1964.

Michelakis, Pantelis. "Greek Lyrics from the Renaissance to the Eighteenth Century." In *The Cambridge Companion to Greek Lyric*, edited by Felix Budelmann, 336–51. Cambridge: Cambridge University Press, 2009.

Minchin, Elizabeth. "Lists and Catalogues in the Homeric Epics." In *Voice into Text: Orality and Literacy in Ancient Greece*, edited by Ian Worthington, 3–20. Leiden: Brill, 1996.

Modrak, Deborah K. W. *Aristotle's Theory of Language and Meaning*. Cambridge: Cambridge University Press, 2001.

Montiglio, Silvia. *Silence in the Land of Logos*. Princeton: Princeton University Press, 2000.

Moreno, Jesús Luque. "Voces. La clasificación de los sonidos en el mundo antiguo: I. Los gramáticos." *Voces* 7 (1996): 9–44.

Murdock, George Peter. "Cross-Language Parallels in Parental Kin Terms." *Anthropological Linguistics* 1.9 (1959): 1–5.

Müri, Walter. "ΣΥΜΒΟΛΟΝ: Wort- und sachgeschichtliche Studie." In *Griechische Studien: Ausgewählte Wortund sachgeschichtliche Forschungen zur Antike*, 1–44. Basel: Friedrich Reinhardt, 1976.

Murray, Penelope and Peter Wilson, eds. *Music and the Muses: The Culture of Mousikē in the Classical Athenian City*. Oxford: Oxford University Press, 2004.

Mustard, Wilfred P. *Classical Echoes in Tennyson*. New York: Macmillan, 1904.

Nagy, Gregory. "Did Sappho and Alcaeus Ever Meet? Symmetries of Myth and Ritual in Performing the Songs of Ancient Lesbos." In *Literatur und Religion I: Wege zu einer mythisch-rituellen Poetik bei den Griechen*, edited by A. Bierl, R. Lämmle, and K. Wesselmann, 211–70. Berlin: Walter de Gruyter, 2007.

——. "Language and Meter." In *A Companion to the Ancient Greek Language*, edited by Egbert J. Bakker, 371–87. Chichester, UK: Wiley-Blackwell, 2010.

——. *Pindar's Homer: The Lyric Possession of an Epic Poet*. Baltimore: Johns Hopkins University Press, 1990.

Nancy, Jean-Luc. *Listening*. Translated by Charlotte Mandell. New York: Fordham University Press, 2007.

Nietzsche, Friedrich. *The Birth of Tragedy and Other Writings*. Edited by Raymond

Geuss and Ronald Speirs. Translated by Ronald Speirs. Cambridge: Cambridge University Press, 1999.

Nooter, Sarah. *When Heroes Sing: Sophocles and the Shifting Soundscape of Tragedy*. Cambridge: Cambridge University Press, 2012.

North, William. "Hearing Voices in Late Antiquity: An Aural Approach to Augustine's *Confessions*." In *The Middle Ages in Texts and Texture: Reflections on Medieval Sources*, edited by Jason Glenn, 7–19. Toronto: University of Toronto Press, 2011.

Nussbaum, Martha C., and Hilary Putnam. "Changing Aristotle's Mind." In *Essays on Aristotle's "De Anima,"* edited by Martha C. Nussbaum and Amélie Oksenberg Rorty, 27–56. Oxford: Oxford University Press, 1992.

O'Brien-Moore, Ainsworth. *Madness in Ancient Literature*. Weimar: R. Wagner Sohn, 1924.

Olson, David R. *The World on Paper: The Conceptual and Cognitive Implications of Writing and Reading*. Cambridge: Cambridge University Press, 1994.

Orlandi, Giovanni. "Metrical and Rhythmical Clausulae in Medieval Latin Prose: Some Aspects and Problems." In *Aspects of the Language of Latin Prose*, edited by Tobias Reinhardt, Michael Lapidge, and J.N. Adams, 395–412. Oxford: Oxford University Press, 2005.

O'Rourke Boyle, Marjorie. "Augustine's Heartbeat: From Time to Eternity." *Viator* 38.1 (2007): 19–43.

Osgood, Josiah. "The *Vox* and *Verba* of an Emperor: Claudius, Seneca and *Le Prince Ideal*." *Classical Journal* 102.4 (2007): 329–53.

O'Sullivan, Neil. *Alcidamas, Aristophanes and the Beginnings of Greek Stylistic Theory*. Stuttgart: Franz Steiner, 1992.

Otto, A. *Die Sprichwörter und sprichwörtlichen Redensarten der Römer: Gesammelt und Erklärt*. Leipzig: B. G. Teubner, 1890.

Ovid. *Metamorfosi*. Vol. 2. Edited by Alessandro Barchiesi and Gianpiero Rosati. Milan: A. Mondadori, 2007.

———. *Metamorphoses*. Edited by R. J. Tarrant. Oxford: Oxford University Press, 2004.

Padel, Ruth. *In and Out of the Mind: Greek Images of the Tragic Self*. Princeton: Princeton University Press, 1992.

———. *Whom Gods Destroy: Elements of Greek and Tragic Madness*. Princeton: Princeton University Press, 1995.

Pavlock, Barbara. *The Image of the Poet in Ovid's "Metamorphoses."* Madison: University of Wisconsin Press, 2009.

Payne, Mark. *The Animal Part: Human and Other Animals in the Poetic Imagination*. Chicago: University of Chicago Press, 2010.

———. "The Understanding Ear: Synaesthesia, Paraesthesia and Talking Animals." In *Synaesthesia and the Ancient Senses*, edited by Shane Butler and Alex Purves, 43–52. Durham, UK: Acumen, 2013.

Pellizer, Ezio, and Maurizio Bettini. *Il mito di Narciso: Immagini e racconti dalla Grecia a oggi*. Turin: G. Einaudi, 2003.

Petrarca, Francesco. *Rime, Trionfi, e Poesie latini*, edited by F. Neri et al. Milan: Riccardo Ricciardi, 1951.

Picker, John M. *Victorian Soundscapes*. Oxford: Oxford University Press, 2003.

Poliziano, Angelo. *Letters*. Translated by Shane Butler. Vol. 1. Cambridge, MA: Harvard University Press, 2006.

Porter, James I. "Des sons qu'on ne peut entendre: Cicéron, les 'kritikoi' et la tradition du sublime dans la critique littéraire." In *Cicéron et Philodème: La polémique en philosophie*, edited by Clara Auvray-Assayas and Daniel Delattre, 315–41. Paris: Éditions Rue d'Ulm, 2001.

———. *Nietzsche and the Philology of the Future*. Stanford: Stanford University Press, 2000.

———. *The Origins of Aesthetic Thought in Ancient Greece: Matter, Sensation, and Experience*. Cambridge: Cambridge University Press, 2010.

———. "Rhetoric, Aesthetics, and the Voice." In *Cambridge Companion to Ancient Rhetoric*, 92–108. Cambridge: Cambridge University Press, 2009.

———. "Saussure and Derrida on the Figure of the Voice." *MLN* 101.4 (1986): 871–94.

Powell, Barry B. *Homer and the Origin of the Greek Alphabet*. Cambridge: Cambridge University Press, 1991.

———. "Homer and Writing." In *A New Companion to Homer*, edited by Ian Morris and Barry B. Powell, 3–32. Leiden: Brill, 1997.

Powell, J. G. F. "Cicero's Style." In *The Cambridge Companion to Cicero*, edited by Catherine E. W Steel. Cambridge: Cambridge University Press, 2013.

Power, Timothy. *The Culture of Kitharōidia*. Washington, DC: Center for Hellenic Studies, 2010.

Price, David H. *Threatening Anthropology: McCarthyism and the FBI's Surveillance of*

Activist Anthropologists. Durham, NC: Duke University Press, 2004.

Primmer, Adolf. "The Function of the *Genera Dicendi* in *De Doctrina Christiana* 4." In "*De Doctrina Christiana*": *A Classic of Western Culture*, edited by Duane W. H. Arnold and Pamela Bright, 87–94. Notre Dame: University of Notre Dame Press, 1995.

Proust, Marcel. *Swann's Way*. Translated by C. K. Scott Moncrieff, Terence Kilmartin, and D. J. Enright. New York: Modern Library, 1992.

Pucci, Pietro. "The Song of the Sirens." In *The Song of the Sirens: Essays on Homer*, 1–9. Lanham, MD: Rowman & Littlefield, 1998.

Reeve, Michael D. "Classical Scholarship." In *The Cambridge Companion to Renaissance Humanism*, edited by Jill Kraye. Cambridge: Cambridge University Press, 1996.

Reinhardt, Tobias. "Epicurus and Lucretius on the Origins of Language." *Classical Quarterly* 58.1 (2008): 127–40.

Rimell, Victoria. *Ovid's Lovers: Desire, Difference and the Poetic Imagination*. Cambridge: Cambridge University Press, 2006.

Rizzo, Silvia. *Ricerche sul latino umanistico*. Rome: Edizioni di Storia e Letteratura, 2002.

Roman, Luke. "Review of (M.) Lowrie, Writing, Performance, and Authority in Augustan Rome." *Classical Review* 61.1 (2011): 118–21.

Ronsard, Pierre de. *Oeuvres complètes*. Edited by Jean Céard, Daniel Ménager, and Michel Simonin. Vol. 1. Paris: Gallimard, 1993.

Rosati, Gianpiero. *Narciso e Pigmalione: Illusione e spettacolo nelle Metamorfosi di Ovidio*. Florence: Sansoni, 1983.

——. "Narciso o l'illusione dissolta (Ovidio, *Metam*. III 339-510)." *Maia* 28 (1976): 83–108.

Rosenmeyer, Patricia A. *The Poetics of Imitation: Anacreon and the Anacreontic Tradition*. Cambridge: Cambridge University Press, 1992.

Ruck, Carl. "Duality and the Madness of Herakles." *Arethusa* 9.1 (1976): 53–75.

Rudy, Jason R. *Electric Meters: Victorian Physiological Poetics*. Athens: Ohio University Press, 2009.

Scharffenberger, Elizabeth W. "*Deinon Eribremetas*: The Sound and Sense of Aeschylus in Aristophanes' *Frogs*." *Classical World* 100.3 (2007): 229–49.

Schenkeveld, D. M. "Studies in the History of Ancient Linguistics IV. Developments in the Study of Ancient Linguistics." *Mnemosyne* 43.3/4 (1990): 289–306.

Schmitz, Thomas. "Narrator and Audience in Philostratus' Lives of the Sophists." In *Philostratus*, edited by Ewen L. Bowie and Jaś Elsner, 49–68. Cambridge: Cambridge University Press, 2009.

Shapiro, Alan. *Re-Fashioning Anakreon in Classical Athens*. Wilhelm Fink, 2012.

Shell, Marc. *The Economy of Literature*. Baltimore: Johns Hopkins University Press, 1978.

Silk, M. S. "Heracles and Greek Tragedy." *Greece & Rome* 32.1 (1985): 1–22.

Smyth, Herbert Weir. *Greek Grammar*. Cambridge, MA: Harvard University Press, 1956.

Sontag, Susan. "Against Interpretation." In *Against Interpretation and Other Essays*, 3–14. New York: Dell, 1967.

Stanford, William Bedell. *Greek Metaphor: Studies in Theory and Practice*. Oxford: Blackwell, 1936.

Starr, Raymond J. "Reading Aloud: *Lectores* and Roman Reading." *Classical Journal* 86.4 (1991): 337–43.

Stearn, William T. *Botanical Latin: History, Grammar, Syntax, Terminology and Vocabulary*. 4th ed. Newton Abbot: David & Charles, 1992.

Stein, Gertrude. *Geography and Plays*. Boston: The Four Seas Company, 1922.

Sterne, Jonathan. *The Audible Past: Cultural Origins of Sound Reproduction*. Durham, NC: Duke University Press, 2003.

Stewart, Susan. *Poetry and the Fate of the Senses*. Chicago: University of Chicago Press, 2002.

Stroh, Valahfridus [Wilfried]. "De uocis definitione quadam Stoica." In *Mousopolos Stephanos: Festschrift für Herwig Görgemanns*, edited by Manuel Baumbach, Helga Köhler, and Adolf Martin Ritter, 443–52. Heidelberg: C. Winter, 1998.

Struck, Peter. *Birth of the Symbol: Ancient Readers at the Limits of Their Texts*. Princeton: Princeton University Press, 2004.

Sullivan, Blair. "The Unwritable Sound of Music: The Origins and Implications of Isidore's Memorial Metaphor." In *The Echo of Music: Essays in Honor of Marie Louise Göllner*, edited by Blair Sullivan. Warren, MI: Harmonie Park Press, 2004.

Svenbro, Jesper. *La parole et le marbre: Aux origines de la poétique grecque*. Lund: Lund University, 1976.

Tarrant, R. J. "The Soldier in the Garden and Other Intruders in Ovid's 'Metamorphoses.'" *Harvard Studies in Classical Philology* 100 (2000): 425–38.

"The New Phonograph." *Scientific American Supplement* 632 (1888): 10096–97.

"The Phonograph." *Harper's Weekly*, March 30, 1878.

Thesaurus Linguae Latinae. Leipzig: Teubner, 1900–.

Tunberg, Terence O. "A Study of *Clausulae* in Selected Works by Lorenzo Valla." *Humanistica Lovaniensia* 41 (1992): 104–33.

Uthemann, Karl-Heinz. "Diatribe (A)." Edited by Hubert Cancik and Helmuth Schneider. *Brill's New Pauly*. Leiden: Brill Online, 2012. http://referenceworks.brillonline.com/entries/brill-s-new-pauly/diatribe-e316870.

Valakas, Kostas. "The Use of the Body by Actors in Tragedy and Satyr-Play." In *Greek and Roman Actors: Aspects of an Ancient Profession*, edited by Pat Easterling and Edith Hall, 69–92. Cambridge: Cambridge University Press, 2002.

Van der Eijk, Philip J. "Aristotle's Psycho-Physiological Account of the Soul-Body Relationship." In *Psyche and Soma: Physicians and Metaphysicians on the Mind-Body Problem from Antiquity to Enlightenment*, edited by John P. Wright and Paul Potter, 57–77. Oxford: Oxford University Press, 2000.

Van Hook, Larue. *The Metaphorical Terminology of Greek Rhetoric and Literary Criticism*. Chicago: University of Chicago Press, 1905.

Vernant, Jean-Pierre, and Pierre Vidal-Naquet. *Myth and Tragedy in Ancient Greece*. New York: Zone Books, 1990.

Vinge, Louise. *The Narcissus Theme in Western European Literature up to the Early 19th Century*. Lund: Gleerups, 1967.

Vollkommer, Rainer. *Herakles in the Art of Classical Greece*. Oxford: Oxford University Committee for Archaeology, 1988.

Walker, Jeffrey. "Pathos and Katharsis in 'Aristotelian' Rhetoric: Some Implications." In *Rereading Aristotle's "Rhetoric,"* edited by Alan G. Gross and Arthur E. Walzer, 74–92. Carbondale: Southern Illinois University Press, 2000.

Watkins, Calvert. *The American Heritage Dictionary of Indo-European Roots*. 3rd ed. Boston: Houghton Mifflin Harcourt, 2011.

———. *How to Kill a Dragon: Aspects of Indo-European Poetics*. New York: Oxford University Press, 1995.

Welch, Katherine E. *The Roman Amphitheatre: From Its Origins to the Colosseum*. Cambridge: Cambridge University Press, 2007.

West, M. L. *Ancient Greek Music*. Oxford: Oxford University Press, 1992.

———. ed. *Carmina Anacreontea*. Leipzig: Teubner, 1984.

White, Charles. *The Life and Times of Little Richard: The Authorised Biography*. 3rd ed. London: Omnibus Press, 2003.

Whitmarsh, Tim. "Greek and Roman in Dialogue: The Pseudo-Lucianic Nero." *Journal of Hellenic Studies* 119 (1999): 142–60.

Wilde, Oscar. *The Complete Works of Oscar Wilde*, vol. 1. Edited by Bobby Fong and Karl Beckson. Oxford: Oxford University Press, 2000.

Wilkinson, L.P. *Golden Latin Artistry*. Cambridge: Cambridge University Press, 1963.

Wills, Jeffrey. *Repetition in Latin Poetry: Figures of Allusion*. Oxford: Clarendon Press, 1996.

Wilson, Peter. "The Aulos in Athens." In *Performance Culture and Athenian Democracy*, edited by Simon Goldhill and Robin Osborne, 58–95. Cambridge: Cambridge University Press, 1999.

——. "Euripides' Tragic Muse." *Illinois Classical Studies* 24–25 (1999–2000): 427–49.

——. "The Musicians among the Actors." In *Greek and Roman Actors: Aspects of an Ancient Profession*, edited by Pat Easterling and Edith Hall, 39–68. Cambridge: Cambridge University Press, 2002.

Winn, James Anderson. *Unsuspected Eloquence: A History of the Relations Between Poetry and Music*. New Haven, Conn: Yale University Press, 1981.

Witt, Ronald G. *In the Footsteps of the Ancients: The Origins of Humanism from Lovato to Bruni*. Leiden: Brill, 2000.

Wollock, Jeffrey L. *The Noblest Animate Motion: Speech, Physiology, and Medicine in Pre-Cartesian Linguistic Thought*. Amsterdam: John Benjamins, 1997.

Woodard, Roger D. *Greek Writing from Knossos to Homer: A Linguistic Interpretation of the Origin of the Greek Alphabet and the Continuity of Ancient Greek Literacy*. New York: Oxford University Press, 1997.

——. "Phoinikēia Grammata." In *A Companion to the Ancient Greek Language*, edited by Egbert J. Bakker, 25–46. Chichester, UK: Wiley-Blackwell, 2010.

Wooten, Cecil W. *Cicero's "Philippics" and Their Demosthenic Model: The Rhetoric of Crisis*. Chapel Hill: University of North Carolina Press, 1983.

Zecchini, Giuseppe. "Polluce e la politica culturale di Commodo." In *L'Onomasticon di Giulio Polluce: Tra lessicografia e antiquaria*, edited by Cinzia Bearzot, Franca Landucci Gattinoni, and Giuseppe Zecchini, 17–26. Milan: Vita e Pensiero, 2007.

Zumthor, Paul. *La poésie et la voix dans la civilisation médiévale*. Paris: Presses Universitaires de France, 1984.

Index

Zone Books series design by Bruce Mau
Typesetting by Meighan Gale
Image placement and production by Julie Fry
Printed and bound by Maple Press